QUESTIONS THAT DEMAND *ANSWERS*

Is the market system ever compatible with social justice? Why are virtually all intellectuals disenchanted with capitalism? Is economics really a science? What lies ahead for Wall Street?

Is the boom-and-bust business cycle really obsolete? Is economic concentration ultimately good or bad? Why is capitalism in difficulties in England and booming in Japan? Does socialism truly work in such "paradises" as Sweden? How are contemporary Marxist critics faring in light of current developments within the capitalistic system?

These are but a few of the key problems attacked by Irving Kristol, Daniel Bell, Peter F. Drucker, Robert L. Heilbroner, Robert M. Solow, Daniel Seligman, M. A. Adelman, R. A. Gordon, Norman Macrae, Martin Bronfenbrenner, Eli Ginzberg, and Raymond Lubitz in—

CAPITALISM TODAY

SIGNET and MENTOR Titles of Related Interest

☐ **THE AFFLUENT SOCIETY by John Kenneth Galbraith.** In his brilliant, famous book, John Kenneth Galbraith argues that the history of material opulence for the masses is relatively brief in the history of mankind and that economic precepts based upon an economy of poverty no longer apply in an affluent society. (#MY1049—$1.25)

☐ **GREAT MYTHS OF ECONOMICS by Don Paarlberg.** Written with engaging wit and simplicity, the author dispels the major economic myths which have been accepted by the layman as gospel. It illuminates the economic principles governing man's behavior and enables the reader to understand the practical workings of the complicated world in which we live. (#MY1094—$1.25)

☐ **THE NEW INDUSTRIAL STATE by John Kenneth Galbraith.** The place of the corporation in the political, economic, and cultural life of the nation. (#Y3637—$1.25)

☐ **IDEAS OF THE GREAT ECONOMISTS by George Soule.** An expert's interpretations of the doctrines of Veblen, Smith, Mill, Marx, Malthus, Keynes, George, Hayek and others. Mr. Soule traces the development of modern economic theories from the strict governmental control of Mercantilism to Adam Smith's arguments for "laissez-faire" and freedom of private enterprise; from Karl Marx's denunciations of the capitalistic order to George Maynard Keynes' support of government investment to prevent depression. Index. (#MT680—75¢)

THE NEW AMERICAN LIBRARY, INC., P.O. Box 999
Bergenfield, New Jersey 07621

Please send me the SIGNET and MENTOR BOOKS I have checked above. I am enclosing $__________(check or money order—no currency or C.O.D.'s). Please include the list price plus 15¢ a copy to cover mailing costs.

Name______________________________

Address______________________________

City____________State________Zip Code________

Allow at least 3 weeks for delivery

CAPITALISM TODAY

Edited by

DANIEL BELL

&

IRVING KRISTOL

A MENTOR BOOK from
NEW AMERICAN LIBRARY
TIMES MIRROR
New York and Scarborough, Ontario
The New English Library Limited, London

Library of Congress Catalog Card Number: 74-150223

This is an authorized reprint of a hardcover edition published by Basic Books, Inc.

SIGNET, SIGNET CLASSICS, SIGNETTE, MENTOR AND PLUME BOOKS are published *in the United States* by
The New American Library, Inc.,
1301 Avenue of the Americas, New York, New York 10019,
in Canada by The New American Library of Canada Limited,
81 Mack Avenue, Scarborough, 704, Ontario,
in the United Kingdom by The New English Library Limited,
Barnard's Inn, Holborn, London, E.C. 1, England.

FIRST PRINTING, DECEMBER, 1971

PRINTED IN THE UNITED STATES OF AMERICA

THE AUTHORS

M. A. ADELMAN is Professor of Economics at Massachusetts Institute of Technology.

DANIEL BELL is Professor of Sociology at Harvard University.

MARTIN BRONFENBRENNER is Professor of Economics at Carnegie-Mellon University and in the fall of 1971 will be Visiting Professor of Economics at Aoyama-Gakuin University in Tokyo.

PETER DRUCKER is a member of the Graduate Faculty of New York University and the author of, among other works, *The Age of Discontinuity* and, most recently, *Technology, Management, and Society*.

ELI GINZBERG is Professor of Economics and Director of the Conservation of Human Resources at Columbia University. He served as consultant to the Ford Foundation in Sweden this year.

R. A. GORDON is Professor of Economics at the University of California at Berkeley.

ROBERT L. HEILBRONER is Professor of Economics at The New School.

IRVING KRISTOL is Professor of Urban Values at New York University.

RAYMOND LUBITZ is Assistant Professor of Economics at Columbia University.

NORMAN MACRAE, Deputy Editor of *The Economist* of London, is the author of *The Neurotic Trillionaire: A Survey of Mr. Nixon's America*.

DANIEL SELIGMAN is Senior Staff Editor at Time, Inc.

ROBERT M. SOLOW is Professor of Economics at Massachusetts Institute of Technology.

INTRODUCTION

MOST HISTORICAL EPOCHS are baptized long after they are buried. Men who lived in the middle ages had no idea that they were living under "feudalism"—any more, of course, than they knew they were living in "the middle ages." Our own era, history-minded to an unprecedented degree, is an exception to this general rule. The term "capitalism" emerged around the middle of the nineteenth century, and it has stuck ever since. But what is even more striking than this act of self-conscious historical description is the fact that this name, "capitalism," is not at all self-congratulatory and is not even neutral. It is a term that, from the beginning, has had a critical edge to it. One might even say, without much exaggeration, that it is mainly a socialist term with an implied adversary intention.

If we have chosen to include that term in the title of this volume, it is because we feel that this self-condemning consciousness of the modern age is as integral a part of it as what are conventionally thought to be the "objective" economic and social realities. This explains why the first two essays are basically essays in cultural history. One cannot understand the important changes that have taken place, and are taking place, in modern society without taking full account of capitalism's uneasy self-consciousness. This self-consciousness is no mere ideological superstructure. It is one of the most fateful and fundamental realities of the system itself.

It is not, of course, the only reality. Institutional change—rapid, profound, and with unanticipated consequences—is another crucial characteristic of modern society, which is as instinctively and organically committed to innovation as previous societies were to stability. The evolution from a social order that celebrated the "producer's ethic" (frugality, industry, sobriety) to one that glorifies the "consumer's ethic" (enjoy now, pay later) was no fortuitous incident in the history of mass psychology. It was made possible by the emergence of an "affluent society," offering the consumer a delicious variety of goods at prices he could, under the new circumstances, afford. And this possibility, in turn, was linked to the steady growth in the productivity of labor as a consequence of technological innovation and mass education, and to the advent of ingenious new modes of financing economic expansion (that is, increasing the productivity of capital). It is possible that had previous generations been given the opportunity to mortgage their future—through installment buying, for example–they would have cheerfully accepted it. Modern society did realize this opportunity—without, however, quite understanding its implications. Hedonism has never been regarded as one of the bourgeois virtues; yet the economics of modern bourgeois society has more and more come to rest on hedonist premises.

A Marxist might say, with benefit of hindsight, that this "contradiction" was inherent in capitalism from the outset. But, in fact, most Marxist predictions were pointed in quite the opposite direction: the "contradictions" of capitalism were supposed to arise out of progressive immiserization, not out of progressive enrichment. And it is a striking characteristic of modern societies that they breed unexpected "contradictions" with a startling fecundity. Not only unexpected, but also frequently unperceived long after they have been firmly established in reality. Thus, critics of capitalism still deplore the competitive scramble for place—"dog eat dog"—that a profit-oriented labor market imposes on the working population. Yet in the United States today, more than one out of four in the labor force works for a non-profit organization (government, private schools and universities, hospitals, philanthropies, foundations, churches, and so on). Forty years ago, most socialists would have thought that only a major political upheaval could engender such a huge expansion of the non-profit sector. But it has happened without anyone paying particular notice, and most Americans today are still unaware that they live in so "mixed" an economy.

In a world that experiences momentous change with such mindlessness, there is bound to be a lot of anxiety beneath the veneer of cheerful expectations. And this subterranean anxiety—which all our poets record, as all our official pronouncements ignore—in turn provokes ideological regression. It is a striking characteristic of our own times that men who only yesterday had fairly coherent, if doctrinaire, ideologies are today vehemently expressing incoherent and bizarre opinions that barely pretend to a purchase on reality. The "end of ideology" has been succeeded by a "greening" of social thought that is itself a signal of severe spiritual crisis.

The essays which follow do not pretend to "solve" this crisis. One can only hope that they will provide some basis for sober thought about it. For reasons of space, many important areas of modern society are ignored. The subject—which is nothing less than the way we live now, and why—is inexhaustible. But the willingness to approach this subject in a calm, self-disciplined spirit of observation and analysis is probably more important than any particular finding or argument.

DANIEL BELL
IRVING KRISTOL

January, 1971

CONTENTS

CAPITALISM ABROAD

1

"When virtue loses all her loveliness" —some reflections on capitalism and "the free society"

IRVING KRISTOL

WHEN we lack the will to see things as they really are, there is nothing so mystifying as the obvious. This is the case, I think, with the new upsurge of radicalism that is now shaking much of Western society to its foundations. We have constructed the most ingenious sociological and psychological theories—as well as a few disingenuously naive ones—to explain this phenomenon. But there is in truth no mystery here. Our youthful rebels are anything but inarticulate; and though they utter a great deal of nonsense, the import of what they are saying is clear enough. What they are saying is that they dislike—to put it mildly—the liberal, individualist, capitalist civilization that stands ready to receive them as citizens. They are rejecting this offer of citizenship and are declaring their desire to see some other kind of civilization replace it.

That most of them do not always put the matter as explicitly or as candidly as this is beside the point. Some of them do, of course; we try to dismiss them as "the lunatic fringe." But the mass of dissident young are not, after all, sufficiently educated to understand the implications of everything they say. Besides, it is so much easier for the less bold among them to insist that what they find outrageous are the defects and shortcomings of the present system. Such shortcomings undeniably exist and are easy polemical marks. And, at the other end, it is so much easier for the adult generations to accept such polemics as representing the sum and substance of their dissatisfaction. It is consoling to think that the turmoil among them is provoked by the extent to which our society falls short of realizing its ideals. But the plain truth is that it is these ideals themselves that are being rejected. Our young radicals are far less dismayed at America's failure to become what it ought to be than they are contemptuous of what it thinks it ought to be. For them, as for Oscar Wilde, it is not the average American who is disgusting; it is the ideal American.

This is why one can make so little impression on them with arguments about how much progress has been made in the past decades, or is being made today, toward racial equality, or abolishing poverty, or fighting pollution, or whatever it is that we conventionally take as a sign of "progress." The obstinacy with which they remain deaf to such "liberal" arguments is not all perverse or irrational, as some would like to think. It arises, rather, out of a perfectly sincere, if often inchoate, animus against the American system itself. This animus stands for a commitment—*to* what, remains to be seen, but *against* what is already only too evident.

Capitalism's three promises

Dissatisfaction with the liberal-capitalist ideal, as distinct from indignation at failures to realize this ideal, are coterminous with the history of capitalism itself. Indeed, the cultural history of the capitalist epoch is not much more than a record of the varying ways such dissatisfaction could be expressed—in poetry, in the novel, in the drama, in painting, and today even in the movies. Nor, again, is there any great mystery why, from the first stirrings of the romantic movement, poets and philosophers have never had much regard for the capitalist civilization in which they lived and worked. But to understand this fully, one must be able to step outside the "progressive" ideology which makes us assume that liberal capitalism is the "natural" state of man toward which humanity has always aspired. There is nothing more natural about capitalist civilization than about many others that have had, or will have, their day. Capitalism represents a sum of human choices about the good life and the good society. These choices inevitably have their associated costs, and after two hundred years the conviction seems to be spreading that the costs have got out of line.

What did capitalism promise? First of all, it promised continued improvement in the material conditions of all its citizens, a promise without precedent in human history. Secondly, it promised an equally unprecedented measure of individual freedom for all of these same citizens. And lastly, it held out the promise that, amidst this prosperity and liberty, the individual could satisfy his instinct for self-perfection—for leading a virtuous life that satisfied the demands of his spirit (or, as one used to say, his soul)—and that the free exercise of such individual virtue would aggregate into a just society.

Now, it is important to realize that, though these aims were in one sense more ambitious than any previously set forth by a political ideology, in another sense they were far more modest. Whereas, as Joseph Cropsey has pointed out, Adam Smith defined "prudence" democratically as "the care of the health, of the fortune, of the rank of the individual," Aristotle had defined that term aristocratically, to mean "the quality of mind concerned with things just and noble and good for man." By this standard, all pre-capitalist systems had been, to one degree or another, Aristotelian: they were interested in creating a high and memorable civilization even if this were shared only by a tiny minority. In contrast, capitalism lowered its sights, but offered its shares in bourgeois civilization to the entire citizenry. Tocqueville, as usual, astutely caught this difference between the aristocratic civilizations of the past and the new liberal capitalism he saw emerging in the United States:

> In aristocratic societies the class that gives the tone to opinion and has the guidance of affairs, being permanently and hereditarily placed above the multitude, naturally conceives a lofty idea of itself and man. It loves to invent for him noble pleasures, to carve out splendid objects for his ambition. Aristocracies often commit very tyrannical and inhuman actions, but they rarely entertain groveling thoughts. . . .
>
> [In democracies, in contrast] there is little energy of character but customs are mild and laws humane. If there are few instances of exalted heroism or of virtues of the highest, brightest, and purest temper, men's

> habits are regular, violence is rare, and cruelty almost unknown. . . . Genius becomes rare, information more diffused. . . . There is less perfection, but more abundance, in all the productions of the arts.

It is because "high culture" inevitably has an aristocratic bias—it would not be "high" if it did not—that, from the beginnings of the capitalist era, it has always felt contempt for the bourgeois mode of existence. That mode of existence purposively depreciated the very issues that were its *raison d'être*. It did so by making them, as no society had ever dared or desired to do, matters of personal taste, according to the prescription of Adam Smith in his *Theory of Moral Sentiments:*

> Though you despise that picture, or that poem, or even that system of philosophy, which I admire, there is little danger of our quarreling upon that account. Neither of us can reasonably be much interested about them. They ought all of them to be matters of great indifference to us both; so that, though our opinions may be opposite, our affections shall be very nearly the same.

In short, an amiable philistinism was inherent in bourgeois society, and this was bound to place its artists and intellectuals in an antagonistic posture toward it. This antagonism was irrepressible—the bourgeois world could not suppress it without violating its own liberal creed; the artists could not refrain from expressing their hostility without denying their most authentic selves. But the conflict could, and was, contained so long as capitalist civilization delivered on its three basic promises. It was only when the third promise, of a virtuous life and a just society, was subverted by the dynamics of capitalism itself, as it strove to fulfill the other two—affluence and liberty—that the bourgeois order came, in the minds of the young especially, to possess a questionable legitimacy.

From bourgeois society to a "free society"

I can think of no better way of indicating the distance that capitalism has travelled from its original ideological origins than by contrasting the most intelligent defender of capitalism today with his predecessors. I refer to Friederich von Hayek, who has as fine and as powerful a mind as is to be found anywhere, and whose *Constitution of Liberty* is one of the most thoughtful works of the last decades. In that book, he offers the following argument against viewing capitalism as a system that incarnates any idea of justice:

> Most people will object not to the bare fact of inequality but to the fact that the differences in reward do not correspond to any recognizable differences in the merit of those who receive them. The answer commonly given to this is that a free society on the whole achieves this kind of justice. This, however, is an indefensible contention if by justice is meant proportionality of reward to moral merit. Any attempt to found the case for freedom on this argument is very damaging to it, since it concedes that material rewards ought to be made to correspond to recognizable merit and then opposes the conclusion that most people will draw from this by an assertion which is untrue. The proper answer is that in a free society it is neither desirable nor practicable that material rewards should be made generally to correspond to what men recognize as merit and that it is an essential characteristic of a free society that an individual's position should not necessarily depend on the views that his fellows hold about the merit he has acquired. . . . A society in which the position of the individual was made to correspond to human ideas of moral merit would therefore be the exact opposite of a free society. It would be a society in which people were rewarded for duty performed instead of for success. . . . But if nobody's knowledge is sufficient to guide all human action, there is also no human being who is competent to reward all efforts according to merit.

This argument is admirable both for its utter candor and for its firm opposition to all those modern authoritarian ideologies, whether rationalist or irrationalist, which give a self-selected elite the right to shape men's lives and fix their destinies according to its preconceived notions of good and evil, merit and demerit. But it is interesting to note what Hayek is doing: he is opposing a *free* society to a *just* society—because, he says, while we know what freedom is, we have no generally accepted knowledge of what justice is. Elsewhere he writes:

> Since they [i.e., differentials in wealth and income] are not the effect of anyone's design or intentions, it is meaningless to describe the manner in which the market distributed the good things of this world among particular people as just or unjust. . . . No test or criteria have been found or can be found by which such rules of "social justice" can be assessed. . . . They would have to be determined by the arbitrary will of the holders of power.

Now, it may be that this is the best possible defense that can be made of a free society. But if this is the case, one can fairly say that "capitalism" is (or was) one thing, and a "free society" another. For capitalism, during the first hundred years or so of its existence, did lay claim to being a just social order, in the meaning later given to that concept by Paul Elmer More: ". . . Such a distribution of power and privilege, and of property as the symbol and instrument of these, as at once will satisfy the distinctions of reason among the superior, and will not outrage the feelings of the inferior." As a matter of fact, capitalism at its apogee saw itself as the most just social order the world has ever witnessed, because it replaced all arbitrary (e.g., inherited) distributions of power, privilege, and property with a distribution that was directly and intimately linked to personal merit—this latter term being inclusive of both personal abilities and personal virtues.

Writing shortly before the Civil War, George Fitzhugh, the most gifted of Southern apologists for slavery, attacked the capitalist North in these terms:

> In a free society none but the selfish virtues are in repute, because none other help a man in the race of competition. In such a society virtue loses all her loveliness, because of her selfish aims. Good men and bad men have the same end in view—self-promotion and self-elevation. . . .

At the time, this accusation was a half-truth. The North was not yet "a free society," in Hayek's sense or Fitzhugh's. It was still in good measure a bourgeois society in which the capitalist mode of existence involved moral self-discipline and had a visible aura of spiritual grace. It was a society in which "success" was indeed seen as having what Hayek has said it ought never to have: a firm connection with "duty performed." It was a society in which Theodore Parker could write of a leading merchant: "He had no uncommon culture of the understanding or the imagination, and of the higher reason still less. But in respect of the *greater faculties*—in respect of conscience, affection, the religious element—he was well born, well bred." In short, it was a society still permeated by the Puritan ethic, the Protestant ethic, the capitalist ethic—call it what you will. It was a society in which it was agreed that there was a strong correlation between certain personal virtues—frugality, industry, sobriety, reliability, piety—and the way in which power, privilege, and property were distributed. And this correlation was taken to be the sign of a just society, not merely of a free one. Samuel Smiles or Horatio Alger would have regarded Professor Hayek's writings as slanderous of his fellow Christians, blasphemous of God, and ultimately subversive of the social order. I am not sure about the first two of these accusations, but I am fairly certain of the validity of the last.

This is not the place to recount the history and eventual degradation of the capitalist ethic in America.[1] Suffice it to say that, with every passing decade, Fitzhugh's charge, that "virtue loses all her loveliness, because of her selfish aims," became more valid. From having been a *capitalist, republican community*, with shared values and a quite unambiguous claim to the title of a just order, the United States became a *free, democratic society* where the will to success and privilege was severed from its moral moorings.

Three current apologia

But can men live in a free society if they have no reason to believe it is also a just society? I do not think so. My reading of history is that, in the same way as men cannot for long tolerate a sense of spiritual meaninglessness in their individual lives, so they cannot for long accept a society in which power, privilege, and property are not distributed according to some morally meaningful criteria. Nor is equality itself any more acceptable than inequality—neither is more "natural" than the other—if

---

[1] See Daniel Bell's essay, "The Cultural Contradictions of Capitalism," in this book for a more detailed analysis of what happened and why.

equality is merely a brute fact rather than a consequence of an ideology or social philosophy. This explains what otherwise seems paradoxical: that small inequalities in capitalist countries can become the source of intense controversy while relatively larger inequalities in socialist or communist countries are blandly overlooked. Thus, those same young radicals who are infuriated by trivial inequalities in the American economic system are quite blind to grosser inequalities in the Cuban system. This is usually taken as evidence of hypocrisy or self-deception. I would say it shows, rather, that people's notions of equality or inequality have extraordinarily little to do with arithmetic and almost everything to do with political philosophy.

I believe that what holds for equality also holds for liberty. People feel free when they subscribe to a prevailing social philosophy; they feel unfree when the prevailing social philosophy is unpersuasive; and the existence of constitutions or laws or judiciaries have precious little to do with these basic feelings. The average working man in nineteenth-century America had far fewer "rights" than his counterpart today; but he was far more likely to boast about his being a free man.

So I conclude, despite Professor Hayek's ingenious analysis, that men cannot accept the historical accidents of the marketplace—seen merely as accidents—as the basis for an enduring and legitimate entitlement to power, privilege, and property. And, in actual fact, Professor Hayek's rationale for modern capitalism is never used outside a small academic enclave; I even suspect it cannot be believed except by those whose minds have been shaped by overlong exposure to scholasticism. Instead, the arguments offered to justify the social structure of capitalism now fall into three main categories:

1) *The Protestant Ethic*—This, however, is now reserved for the lower socioeconomic levels. It is still believed, and it is still reasonable to believe, that worldly success among the working class, lower-middle class, and even middle class has a definite connection with personal virtues such as diligence, rectitude, sobriety, honest ambition, etc., etc. And, so far as I can see, the connection is not only credible but demonstrable. It does seem that the traditional bourgeois virtues are efficacious among these classes—at least, it is rare to find successful men emerging from these classes who do not to a significant degree exemplify them. But no one seriously claims that these

traditional virtues will open the corridors of corporate power to anyone, or that the men who now occupy the executive suites are—or even aspire to be—models of bourgeois virtue.

2) *The Darwinian Ethic*—This is to be found mainly among small businessmen who are fond of thinking that their "making it" is to be explained as "the survival of the fittest." They are frequently quite right, of course, in believing the metaphor appropriate to their condition and to the ways in which they achieved it. But it is preposterous to think that the mass of men will ever accept as legitimate a social order formed in accordance with the laws of the jungle. Men may be animals, but they are political animals—and, what comes to not such a different thing, moral animals too. The fact that for several decades after the Civil War, the Darwinian ethic, as popularized by Herbert Spencer, could be taken seriously by so many social theorists represents one of the most bizarre and sordid episodes in American intellectual history. It could not last; and did not.

3) *The Technocratic Ethic*—This is the most prevalent justification of corporate capitalism today, and finds expression in an insistence on "performance." Those who occupy the seats of corporate power, and enjoy the prerogatives and privileges thereof, are said to acquire legitimacy by their superior ability to achieve superior "performance"—in economic growth, managerial efficiency, technological innovation. In a sense, what is claimed is that these men are accomplishing social tasks, and fulfilling social responsibilities, in an especially efficacious way.

There are, however, two fatal flaws in this argument. First, if one defines "performance" in a strictly limited and measurable sense, then one is applying a test that any ruling class is bound, on fairly frequent occasions, to fail. Life has its ups and downs; so do history and economics; and men who can only claim legitimacy *via* performance are going to have to spend an awful lot of time and energy explaining why things are not going as well as they ought to. Such repeated, defensive apologias, in the end, will be hollow and unconvincing. Indeed, the very concept of "legitimacy," in its historical usages, is supposed to take account of and make allowances for all those rough passages a society will have to navigate. If the landed gentry of

Britain during those centuries of its dominance, or the business class in the United States during the first century and a half of our national history, had insisted that it be judged by performance alone, it would have flunked out of history. So would every other ruling class that ever existed.

Secondly, if one tries to avoid this dilemma by giving the term "performance" a broader and larger meaning, then one inevitably finds oneself passing beyond the boundaries of bourgeois propriety. It is one thing to say with Samuel Johnson that men honestly engaged in business are doing the least mischief that men are capable of; it is quite another thing to assert that they are doing the greatest good—this is only too patently untrue. For the achievement of the greatest good, more than successful performance in business is necessary. Witness how vulnerable our corporate managers are to accusations that they are befouling our environment. What these accusations really add up to is the statement that the business system in the United States does not create a beautiful, refined, gracious, and tranquil civilization. To which our corporate leaders are replying: "Oh, we can perform that mission too—just give us time." But there is no good reason to think they can accomplish this noncapitalist mission; nor is there any reason to believe that they have any proper entitlement even to try.

"Participation" or leadership?

It is, I think, because of the decline of the bourgeois ethic, and the consequent drainage of legitimacy out of the business system, that the issue of "participation" has emerged with such urgency during these past years. It is a common error to take this word at its face value—to assume that, in our organized and bureaucratized society, the average person is more isolated, alienated, or powerless than ever before, and that the proper remedy is to open new avenues of "participation." We are then perplexed when, the avenues having been open, we find so little traffic passing through. We give college students the right to representation on all sorts of committees—and then discover they never bother to come to meetings. We create new popularly-elected "community" organizations in the ghettos—and then discover that ghetto residents won't come out to vote. We decentralize New York City's school system—only to discover that the populace is singularly uninterested in local school board elections.

I doubt very much that the average American is actually more isolated or powerless today than in the past. The few serious studies that have been made on this subject indicate that we have highly romanticized notions of the past—of the degree to which ordinary people were ever involved in community activities—and highly apocalyptic notions of the present. If one takes membership in civic-minded organizations as a criterion, people are unquestionably more "involved" today than ever before in our history. Maybe that's not such a good criterion; but it is a revealing aspect of this whole problem that those who make large statements on this matter rarely give us any workable or testable criteria at all.

But I would not deny that more people, even if more specifically "involved" than ever before, also feel more "alienated" in a general way. And this, I would suggest, is because the institutions of our society have lost their vital connection with the values which are supposed to govern the private lives of our citizenry. They no longer exemplify these values; they no longer magnify them; they no longer reassuringly sustain them. When it is said that the institutions of our society have become appallingly "impersonal," I take this to mean that they have lost any shape that is congruent with the private moral codes which presumably govern individual life. (That presumption, of course, may be factually weak; but it is nonetheless efficacious so long as people hold it.) The "outside" of our social life has ceased being harmonious with the "inside"—the mode of distribution of power, privilege, and property, and hence the very principle of authority, no longer "makes sense" to the bewildered citizen. And when institutions cease to "make sense" in this way, all the familiar criteria of success or failure become utterly irrelevant.

As I see it, then, the demand for "participation" is best appreciated as a demand for authority—for leadership that holds the promise of reconciling the inner and outer worlds of the citizen. So far from its being a hopeful reawakening of the democratic spirit, it signifies a hunger for authority that leads toward some kind of plebiscitary democracy at best, and is in any case not easy to reconcile with liberal democracy as we traditionally have known it. I find it instructive that such old-fashioned populists as Hubert Humphrey and Edmund Muskie, whose notions of "participation" are both liberal and traditional, fail to catch the imagination of our dissidents in the way that Robert Kennedy did. The late Senator Kennedy was very much a leader—one can imagine Humphrey or Muskie participating in an old-fashioned town meeting, one can only envision Kennedy dominating a town rally. One can also envision those who "participated" in such a rally feeling that they had achieved a kind of "representation" previously denied them.

A case of regression

For a system of liberal, representative government to work, free elections are not enough. The results of the political process and of the exercise of individual freedom—the distribution of power, privilege, and property—must also be seen as in some profound sense expressive of the values that govern the lives of individuals. An idea of self-government, if it is to be viable, must encompass both the private and public sectors. If it does not—if the principles that organize public life seem to have little relation to those that shape private lives—you have "alienation," and *anomie,* and a melting away of established principles of authority.

Milton Friedman, arguing in favor of Hayek's extreme libertarian position, has written that the free man "recognizes no national purpose except as it is the consensus of the purposes for which the citizens severally strive." If he is using the term "consensus" seriously, then he must be assuming that there is a strong homogeneity of values among the citizenry, and that these values give a certain corresponding shape to the various institutions of society, political and economic. Were that the case, then it is indeed true that a "national purpose" arises automatically and organically out of the social order itself. Something like this did happen when liberal capitalism was in its prime, vigorous and self-confident. But is that our condition today? I think not—just as I think Mr. Friedman doesn't really mean "consensus" but rather the mere aggregation of selfish aims. In such a blind and accidental arithmetic, the sum floats free from the addenda, and its legitimacy is infinitely questionable.

The inner spiritual chaos of the times, so powerfully created by the dynamics of capitalism itself, is such as to make nihilism an easy temptation. A "free society" in

Hayek's sense gives birth in massive numbers to "free spirits"—emptied of moral substance but still driven by primordial moral aspirations. Such people are capable of the most irrational actions. Indeed, it is my impression that, under the strain of modern life, whole classes of our population—and the educated classes most of all—are entering what can only be called, in the strictly clinical sense, a phase of infantile regression. With every passing year, public discourse becomes sillier and more petulant, while human emotions become, apparently, more ungovernable. Some of our most intelligent university professors are now loudly saying things that, had they been uttered by one of their students twenty years ago, would have called forth gentle and urbane reproof.

The reforming spirit and the conservative ideal

And yet, if the situation of liberal capitalism today seems so precarious, it is likely nevertheless to survive for a long while, if only because the modern era has failed to come up with any plausible alternatives. Socialism, communism, and fascism have all turned out to be either utopian illusions or sordid frauds. So we shall have time—though not an endless amount of it, for we have already wasted a great deal. We are today in a situation not very different from that described by Herbert Croly in *The Promise of American Life* (1912):

> The substance of our national Promise has consisted . . . of an improving popular economic condition, guaranteed by democratic political institutions, and resulting in moral and social amelioration. These manifold benefits were to be obtained merely by liberating the enlightened self-enterprise of the American people. . . . The fulfillment of the American Promise was considered inevitable because it was based upon a combination of self-interest and the natural goodness of human nature. On the other hand, if the fulfillment of our national Promise can no longer be considered inevitable, if it must be considered as equivalent to a conscious national purpose instead of an inexorable national destiny, the implication necessarily is that the trust reposed in individual self-interest has been in some measure betrayed. No pre-established harmony can then exist between the free and abundant satisfac-

tion of private needs and the accomplishment of a morally and socially desirable result.

Croly is not much read these days. He was a liberal reformer with essentially conservative goals. So was Matthew Arnold, fifty years earlier—and he isn't much read these days, either. Neither of them can pass into the conventional anthologies of liberal or conservative thought. I think this is a sad commentary on the ideological barrenness of the liberal and conservative creeds. I also think it is a great pity. For if our private and public worlds are ever again, in our lifetimes, to have a congenial relationship—if virtue is to regain her lost loveliness—then some such combination of the reforming spirit with the conservative ideal seems to me to be what is most desperately wanted.

I use the word "conservative" advisedly. Though the discontents of our civilization express themselves in the rhetoric of "liberation" and "equality," one can detect beneath the surface an acute yearning for order and stability—but a legitimate order, of course, and a legitimized stability. In this connection, I find the increasing skepticism as to the benefits of economic growth and technological innovation most suggestive. Such skepticism has been characteristic of conservative critics of liberal capitalism since the beginning of the nineteenth century. One finds it in Coleridge, Carlyle, and Newman—in all those who found it impossible to acquiesce in a "progressive" notion of human history or social evolution. Our dissidents today may think they are exceedingly progressive; but no one who puts greater emphasis on "the quality of life" than on "mere" material enrichment can properly be placed in that category. For the idea of progress in the modern era has always signified that the quality of life would inevitably be improved by material enrichment. To doubt this is to doubt the political metaphysics of modernity and to start the long trek back to pre-modern political philosophy—Plato, Aristotle, Thomas Aquinas, Hooker, Calvin, etc. It seems to me that this trip is quite necessary. Perhaps there we shall discover some of those elements that are most desperately needed by the spiritually impoverished civilization that we have constructed on what once seemed to be sturdy bourgeois foundations.

2

The cultural contradictions of capitalism

DANIEL BELL

THE relationship between a civilization's socioeconomic structure and its culture is perhaps the most complicated of all problems for the sociologist. A nineteenth century tradition, one deeply impregnated with Marxist conceptions, held that changes in social structure determined man's imaginative reach. An earlier vision of man—as *homo pictor,* the symbol-producing animal, rather than as *homo faber,* the tool-making animal—saw him as a creature uniquely able to prefigure what he would later "objectify" or construct in reality. It thus ascribed to the realm of culture the initiative for change. Whatever the truth of these older arguments about the past, today culture has clearly become supreme; what is played out in the imagination of the artist foreshadows, however dimly, the social reality of tomorrow.

Culture has become supreme for two complementary reasons. First, culture has become the most dynamic com-

---

This essay is an abridgement of two chapters from a book, tentatively titled, The Divided Society: On the Disjunction of Culture and Social Structure, *to be published by Basic Books in 1971. In this version, much of the supporting argument and footnote citation necessarily have been heavily reduced. The book expands the historical argument and tries to delineate a sociological theory of culture. This book was written during my tenure as a Visiting Scholar at the Russell Sage Foundation.—DB*

ponent of our civilization, outreaching the dynamism of technology itself. There is now in art—as there has increasingly been for the past hundred years—a dominant impulse towards the new and the original, a self-conscious search for future forms and sensations, so that *the idea* of change and novelty overshadows the dimensions of actual change. And secondly, there has come about, in the last fifty years or so, a legitimation of this cultural impulse. Society now accepts this role for the imagination, rather than—as in the past—seeing it as establishing a norm and affirming a moral-philosophic tradition against which the new could be measured and (more often than not) censured. Indeed, society has done more than passively accept —it has provided a market which eagerly gobbles up the new, because it believes it to be superior in value to all older forms. Thus, our culture has an unprecedented mission: it is an official, ceaseless searching for a new sensibility.

It is true, of course, that the idea of change dominates the modern economy and modern technology as well. But changes in the economy and technology are constrained by available resources and financial cost. In politics, too, innovation is constrained by existing institutional structures, by the veto power of contending groups, and to some extent by tradition. But the changes in expressive symbols and forms, difficult as it may be for the mass of people to absorb them readily, meet no resistance in the realm of culture itself.

What is singular about this "tradition of the new" (as Harold Rosenberg has called it) is that it allows art to be unfettered, to break down all genres and to explore all modes of experience and sensation. Fantasy today has few costs (is *anything* deemed bizarre or opprobrious today?) other than the risk of individual madness. And even madness, in the writings of such social theorists as Michel Foucault and R. D. Laing, is now conceived to be a superior form of truth! The new sensibilities, and the new styles of behavior associated with them, are created by small coteries which are devoted to exploring the new; and because the new has value in and of itself, and meets with so little resistance, the new sensibility and its behavior-styles diffuse rapidly, transforming the thinking and actions of larger masses of people.

Along with this emphasis on the new has come the ideology, self-consciously accepted by the artist, that art will lead the way, will serve as the *avant-garde*. Now the very idea of an *avant-garde*—an advance assault team—indicates that modern art or culture would never permit itself to serve as a "reflection" of an underlying social structure, but rather would open the way to something radically new. In fact, as we shall see, the very idea of an *avant-garde,* once its legitimacy is accepted, serves to institutionalize the primacy of culture in the fields of manners, morals, and ultimately politics.

The first major formulation of this conception of the *avant-garde* was by the man who, ironically, has come to serve as the symbol of technocratic rule, Henri de Saint-Simon. For all his vision of the engineer as the driving force of the new society, Saint-Simon knew that men were in want of inspiration, that Christianity itself was worn out, and that a new cult was needed. He found this new cult in the cult of art itself. The artist would reveal to society the glorious future, exciting men with the prospect of a new civilization. In a dialogue between an artist and a scientist Saint-Simon gave the phrase its modern *cultural*—rather than its earlier military—meaning:

> It is we, artists, who will serve you as *avant-garde:* the power of the arts is in fact most immediate and most rapid. When we wish to spread new ideas among men, we inscribe them on marble or on canvas; . . . and in that way above all we exert an electric and victorious influence. We address ourselves to the imagination and to the sentiments of mankind, we should therefore always exercise the liveliest and the most decisive action. . . .
>
> What a most beautiful destiny for the arts, that of exercising over society a positive power, a true priestly function, and of marching forcefully in the van of all the intellectual faculties in the epoch of their greatest development! This is the duty of artists and their mission. . . .

The commonplace observation that today there is no longer a significant *avant-garde*—that there is no longer a radical tension between a new art which shocks and a society that is shocked—merely signifies that the *avant-garde* has won its victory. A society given over entirely to innovation, in the joyful acceptance of change, has in fact institutionalized an *avant-garde* and charged it—perhaps to its own eventual dismay—with constantly turning up something new. In effect, "culture" has been given a blank check, and its primacy in generating social change has been firmly acknowledged.

Discretionary social behavior

This changeover creates a new and peculiar set of historic tensions in the society. The social structure today is ruled by an economic principle of rationality, defined in terms of efficiency in the allocation of resources; the culture, in contrast, is prodigal, promiscuous, dominated by an antirational, anti-intellectual temper. The character structure inherited from the nineteenth century—with its emphasis on self-discipline, delayed gratification, restraint—is still relevant to the demands of the social structure; but it clashes sharply with the culture, where such bourgeois values have been completely rejected—in part, as we shall see, and paradoxically, because of the workings of the capitalist economic system itself.

Our prevailing social theories are utterly confounded by the new currents in our culture. Pitirim Sorokin's idea of the modern world having a "sensate mentality"—empirical, materialistic, technological—is contradicted in good part by the rise of hallucinogenic and psychedelic experience, the search for community, and the rejection of "material possession" by a significant section of the new culture-bearing elites. Rather than conform to Max Weber's theory of the special appropriateness of rational forms of thought and behavior to twentieth-century society, we see in all the arts a breakup of rational cosmology: of foreground and background in painting; of sequence, beginning, middle, and end in narrative; of melody and harmonic tonalities in music. Against the classical theories of distinguishable disciplines, we find the breakup of genres and an emphasis on "total environments," i.e., so-called "anti-art" movements which erase the distinction between art and everyday experience. And contrary to Marx' idea of culture "reflecting" an economy, integrally tied to it through the exchange process, two distinct and extraordinary changes are taking place. Art has become increasingly autonomous, making the artist a powerful taste-maker in his own right; the "social location" of the individual (his social class or other position) no longer determines his life-style and his values.

These changes—the search for new esthetic experience,

the break-up of formal genres, and the detachment of life-styles from a fixed social base—have become most evident in the last decade, and create the most perplexing problems for social analysis. As a discipline, sociology assumes that variations in behavior of persons or groups in the society are attributable to their class or some other strategic position in the social structure, and that individuals so differentially placed will vary systematically in their interests, attitudes, and behavior on the basis of distinct social attributes: e.g., common age, sex, occupation, religion, urban-rural location, and the like. The presumption is that these attributes cluster in specific ways—usually identified in social-class terms—so that voting behavior, buying habits, child-rearing vary systematically on a class basis and are predictable.

For the majority of the society, and for many aspects of social life (e.g., voting), this general proposition may still hold true. But it is increasingly evident that for a significant proportion of the population the relation of social position to cultural style—particularly if one thinks in gross dimensions such as working class, middle class, and upper class—no longer holds. The question of who will use drugs, engage in orgies and wife-swapping, become an open homosexual, use obscenity as a political style, enjoy "happenings" and underground movies, is not easily related to the "standard variables" of sociological discourse. Age and education may be more relevant discriminators; but in the expansion of mass higher education, even education alone is no longer an easy predictor of behavior. One finds many children of upper-middle-class families joyfully embracing what they think is the "freedom" of working-class or black, lower-class life-styles—and others who do not. There is a significant levelling in patterns of child-training, which was one of the major indicators of different class styles in the past.

Just as in the economy the growth of what economists call *discretionary income*—income above that necessary for the fulfillment of basic needs—allowed individuals to choose many varied items to exemplify a different consumption style (swimming pools, boats, travel), so the expansion of higher education and the extension of a permissive social atmosphere has widened the scope of *discretionary social behavior*. The more idiosyncratic aspects of personal experience and life-history—personality attributes, or somatic body-type constitution, positive or negative experience with parents, experience with peers—

become increasingly more important than patterned social attributes in shaping a life-style for a person. As the traditional class structure dissolves, more and more individuals want to be identified, not by their occupational base (in the Marxist sense), but by their cultural tastes and life-styles.

The artist makes the audience

A change has been taking place, as well, in the relation of the artist to the public. The familiar image, a product of nineteenth-century romanticism, was that of a coterie of artists, engaged in difficult experimental work to which the smug middle-class audience responded with scorn and outrage. This was the fate of the Impressionist painters, who appeared first in the *Salon des Refusés* (1863) to emphasize their own disgust with the regnant taste and who had to wait twenty years for the *Salon des Indépendants* for the same freedom to exhibit. The *avant-garde* artist identified this rejection with freedom, and he depended on such tension with the audience to articulate his own work. This well-known pattern came to be regarded as a congenital condition of modern art. But as James Ackerman writes, "within the last decade [this pattern] was broken by one of history's most abrupt and radical changes in the relationship of art and its public . . . the new era became recognizable first in the ultimate reception of the work of the New York School of artists in the mid- and late 1950's.[1] Jackson Pollock, Willem de Kooning, Franz Kline, Mark Rothko, Barnett Newman, Robert Motherwell, David Smith, the men responsible for what Clement Greenberg called "abstract expressionism" (and Harold Rosenberg "action painting"), were preoccupied with problems of structure and medium—breaking away from the easel, using paint itself as a subject for art, involving the person of the artist in the painting—of a special and esoteric nature outside the experience of the layman. Professor Ackerman observes that "Their art was so difficult to approach that even the majority of approving professional critics missed the mark and praised it for irrelevant reasons." In fact, the immediate incredulous

[1] James Ackerman, "The Demise of the *Avant Garde*: Notes on the Sociology of Recent American Art," *Comparative Studies in Society and History*, Vol. 11, No. 4, October 1969, pp. 371-384, esp. p. 378.

public response was to call it a sham. But *within half a decade* the major figures in the school had been acclaimed, and their paintings dominated the museums and the galleries. Their conceptions of art now set the taste for the public.

Perhaps the change in this case is not as abrupt as Professor Ackerman makes it seem. There had been earlier and similar changes in the role of "difficult" art, in Paris decades earlier, when Picasso and Matisse began to shape public taste. But the general point stands. The middle-class audience, or even the buyer alone, no longer controls art. In painting, in film (perhaps less so in advanced music), the artist, and usually the *avant-garde* artist, now dominates the cultural scene. It is he who swiftly shapes the audience and the market, rather than being shaped by them.

The "adversary culture"

This change is related, I believe, to the dissociation of social location and cultural style. Ackerman also writes:

> If one's position in society implies no determinate base of judgment in areas outside one's competence, one has a choice between having no opinion or accepting the opinion of the expert, and the most available expert is the professional manufacturer of opinion. The altered response to the arts is, I believe, a product of public deference to museums, commercial galleries, and the news media.

Whether there is now a general habit of "trusting the experts" is debatable. In politics, there has been a notable populist reaction against the expert or technocrat. But the situation in art is different. Here we see, not the victory of the expert, but of "culture" itself. The culture of the past hundred years, that of the "modern movement," has triumphed over a society that in its social structure—economics, technology, and occupational bases—remains bourgeois. The culture has become detached, and self-determining. Yet with all that, the culture (as exemplified in the modern movement) feels itself under attack—does not understand or accept its victory—and remains, as Lionel Trilling has called it, an "adversary culture."

"Any historian of the literature of the modern age,"

Trilling writes, "will take virtually for granted the adversary intention, the actual subversive intention, that characterizes modern writing—he will perceive its clear purpose of detaching the reader from the habits of thought and feeling that the larger culture imposes, of giving him a ground and a vantage point from which to judge and condemn, and perhaps revise, the culture that has produced him."[2]

The legend of modernism is that of the free creative spirit at war with the bourgeoisie. Whatever the truth of such a view when, say, Whistler was accused of having "flung a pot of paint in the public's face," in our time the idea is a caricature. Who in the world today, especially in the world of culture, defends the bourgeoisie? Yet in the domain of those who think themselves serious about culture, and of their widespread and trailing *epigoni,* the legend of the free creative spirit now at war, no longer merely with bourgeois society, but with "civilization" or "repressive tolerance" or some other agency that curtails "freedom," still sustains an adversary culture.

The impulses of that artistic and intellectual culture have not changed from that of seventy years ago. In terms of programmatic vigor and technical innovation, it reached its apogee in the first quarter of this century (in the work of Eliot, Pound, Proust, Joyce, Picasso, Braque, Schoenberg, Webern). But, as with any successful "movement," what starts out with small coteries begins to diffuse throughout the society. While there is no longer anything intrinsically novel in what is produced, these ideas appeal now to a larger and larger group in the society—so much so, that it has become a reigning ideology and the flag of a dominant cultural class.

The adversary culture has come to dominate the social order, and this is why the hierophants of the culture—the painters, the writers, the filmmakers—now dominate the audience, rather than vice-versa. And, indeed, the subscribers to this adversary culture are sufficiently numerous to form a distinct cultural class. In numbers, compared to the society as a whole, the membership in this class is not large. No statistical estimates are possible, and the figure could vary from a few hundred thousand to a couple of million. But size alone is meaningless for, compared to the past, three extraordinary changes are evident.

First, in size, there has been an evident change of scale.

[2] Lionel Trilling, *Beyond Culture* (New York, 1965), p. xiii.

Even though tiny in comparison with the numbers of the total society, the present size is large enough for these individuals no longer to be outcasts, or a bohemian enclave, in the society. They function institutionally as a group, bound by a consciousness of kind.

Second, while minority life-styles and cultures have often conflicted with those of the majority, what is striking today is that the *majority* has no intellectually respectable culture of its own—no major figures in literature (the best is James Gould Cozzens), painting (except, perhaps, Andrew Wyeth), or poetry—to counterpose to the adversary culture. In this case, bourgeois culture has been shattered.

Third, and perhaps most important: the protagonists of the adversary culture, despite their sincere and avowed subversive intentions, do substantially influence, if not dominate, the cultural establishments today—the publishing houses, museums, galleries; the major news, picture, and cultural weeklies and monthlies; the theatre, film, and the universities.

Today, each new generation, starting off at the benchmarks attained by the adversary culture of their cultural parents, declares in sweeping fashion that the status quo represents a state of absolute repression, so that, in a widening gyre, new and fresh assaults on the social structure are mounted. This, I believe, has been happening in the last two decades.

II

The historic process that I have been sketching has deep roots in the past. It has remarkable cultural drive and continuity. Much of that drive was obscured in the 1950's which was, essentially, a decade of conservatism and cultural bewilderment. Yet in retrospect, it seems clear that the 1950's was an aberrant decade—and that in the 1960's, a radicalism endemic to the society had resumed its historic drive.

Politically, the 1950's was a period of disillusionment. It witnessed the final rupture of the intellectuals from Stalinism, a shattering of the belief that the Soviet Union was "progressive" merely because it called itself socialist. The crucial events were Khrushchev's secret speech before the 22nd Congress of the Russian Communist Party, in February 1956, when he admitted Stalin's brutal murder of former Communist leaders; the burst of political independence in Poland following these revelations; and, more spectacularly, the Hungarian Revolution, which was put down by Soviet tanks and ended in the murder of the Hungarian Communist Party leader, Ferenc Nagy. A number of sociologists—Raymond Aron, Edward Shils, S. M. Lipset, and myself—thus came to view the 1950's as characterized by an "end of ideology." By this we meant that the older political ideas of the radical movement had become exhausted and no longer had the power to compel allegiance or passion among the intelligentsia.[3]

But, although there was a widespread disillusionment with the chiliastic promises of political radicalism, there was almost no positive viewpoint to take its place. The welfare state and the mixed economy were not the sort of goals that could capture the passions of the intelligentsia. Moreover, even if radical political hopes were momentarily shattered, the basic cultural stance remained the same: the rejection of bourgeois values. Indeed, *the continuity of radicalism in the 1950's was possible not through politics but through the culture.*

The experience of the 1940's had traumatized the intelligentsia of the 1950's, and reflection on that decade determined their cultural concerns. The pervasive cultural theme of the era was the depersonalization of the individual and the atomization of society. World War II was horrible, of course. But war, even the mass bombing of cities, had been prefigured in the imagination and, curiously, once something has been imagined, it loses some of its capacity to arouse complete indignation or fear. But concentration camps enfolding tens of millions, and death camps that processed millions of individuals through a slaughterhouse like cattle had never been imagined.

The culture of the 1950's—the writers who were read and studied as exemplars of the contemporary spirit—reflected that incomprehension of totalitarian terror. The primary literary figure was Franz Kafka, whose novels and stories, written thirty years before, were found to

---

[3] I should point out that the analysis of the "end of ideology" did *not* assume that all social conflict had ended and that the intelligentsia would foreswear the search for new ideologies. In fact, as I wrote in 1959: "The young intellectual is unhappy because the 'middle way' is for the middle-aged, not for him; it is without passion and is deadening. . . . In the search for a 'cause' there is a deep desperate, almost pathetic anger," etc. See *The End of Ideology* (Free Press paperback), p. 404.

have anticipated that dense, bureaucratic world where justice could not be located and where the torture machine inflicted a horrible death on its victims. The writings of Kierkegaard were "discovered," perhaps because he counselled that no rational belief in ultimate meanings was possible, only the leap of faith. The neo-orthodox theology of Barth and Niebuhr was pessimistic about man's ability to transcend the sinfulness inherent in human pride. Simone Weil's writings dealt with the desperate search for grace. Camus scrutinized the moral paradoxes of political action. In the "theatre of the absurd," Ionesco wrote plays in which objects, like *The Chairs,* came to have a life of their own, as if the reified things of the world had actually drawn the spirit out of man and taken over his will. In the theatre of silence, such as Beckett's *Waiting for Godot,* the confusions of time and self were played out in a minimal rectangle of reality.

The sociology of the 1950's was similarly concerned with the theory of "the mass society" and the rediscovery of "alienation." The theory of the mass society saw in the modern world the shattering of the traditional primary-group ties of family and local community; it saw traditional orders replaced by the "mass," in which each person lived in atomistic or anomic fashion. The rediscovery of alienation—and it was a *re*discovery, for though it has been associated with Marxism, the first generation of Marxist writers (Kautsky, Plekhanov, Lenin) had never used the term—had a double source. On the one hand it was associated, principally through the writings of Max Weber, with the sense of powerlessness that individuals felt in the society. Marx' emphasis on the worker "separated" from the means of production became, in Weber's perspective, one special case of a universal trend in which the modern soldiers are separated from the means of violence, the scientist from the means of inquiry, and the civil servant from the means of administration. On the other hand, it was a theme put forward by Marxist revisionists, principally the post-Stalinist generation, which hoped to find the sources of a new humanism in Marx' early writings, mainly the *Economic-Philosophical Manuscripts.* In both instances, the theory of mass society and the theme of alienation, what was involved were critical cultural judgments on the quality of life in modern society.

On a more mundane level, the most popular book of sociology in the 1950's was David Riesman's *The Lonely Crowd,* which described a major change in character structure in contemporary society—from an individual who was self-disciplined and self-motivated (in short, the historic bourgeois man) to one who was responsive primarily to his peer group and the pressure of "others." The very title of the book conveyed a judgment about the change. Similarly, the prototypical book of the emerging youth culture in the 1950's was J. D. Salinger's *The Catcher in the Rye,* whose narrator, Holden Caulfield, epitomized a new kind of almost "autistic" generation. The "beats," led by Allen Ginsberg and Jack Kerouac, harbingers of the youth movement of the 1960's, had already "dropped out" of the society.

In short, though political ideas had become exhausted—and political life was dominated by the threat of a foreign communist foe—the cultural intelligentsia brooded on themes of despair, *anomie,* and alienation—themes which were to achieve a political incarnation in the 1960's.

The "middlebrows" of the 1950's

The affluence of middle-class America in the 1950's had its counterpart in a widespread "middlebrow" culture. The term itself reflected the new style of cultural criticism. In effect "culture," as it came to be conceived in the mass middle-class magazines, was not a discussion of serious works of art but a style of life that was organized and "consumed." Following suit, cultural criticism became a snob's game, played by advertising men, magazine illustrators, home decorators, women's magazine editors, and East Side homosexuals as one more fashionable amusement. The game of high-low-and-middle became *demode* once the middlebrows caught on—to be quickly replaced by the new game of "in"-and-"out." To be "in" meant to be well ahead of the crowd in fashion, or, perversely, to like what the vulgar masses liked (the New York *Daily News,* fast-paced grade-B movie thrillers, big popular jazzy dance halls), rather than what the pretentious middle classes liked. When in-and-out was replaced by "camp," the game was the same, except that fashion had become high fashion.

But even though cultural criticism became a game, it was also a serious problem for the intellectual, who was now invited to play a role in a culture he had always mocked. The writers for *Partisan Review* now came to dominate *The New Yorker,* a magazine that was scorned in the 1930's and 1940's. Writers for *Commentary* were invited to write in the *Sunday Times* magazine section. Even the *Saturday Evening Post* began running articles on "Adventures of the Mind," by such writers and critics as Randall Jarrell and Clement Greenberg. Many of the radical writers felt that the mass media invited them in order to provide prestige for the mass magazines; and an even more sinister motive of the "taming" of radical criticism altogether was suspected. What was not realized was that society itself had lost its cultural moorings.

The relationship of the serious critic and intellectual to the burgeoning mass culture of the 1950's became a discrete problem in itself and the source of many a lengthy essay and symposium. The fundamental response of the radical intellectual was a wide-ranging attack on middle-class culture. For the serious critic, the "real" enemy, the worst *kitsch*, was not the vast sea of trash but middlebrow culture; or, as Dwight Macdonald labeled it, "Midcult." In "Masscult," Macdonald writes, "the trick is plain—to please the crowd by any means. But Midcult has it both ways: it pretends to respect the standards of High Culture while in fact it waters them down and vulgarizes them."[4] To critics like Macdonald, the special danger of "Midcult" was that, in the upgrading of American taste and standards, the lines between high culture and Midcult become blurred, and Midcult standards, precisely because they seem to advance culture, would predominate. "We are now in a more sophisticated period. . . . Since 1900, American culture has moved culturally in a direction that on the whole appears to be up," Mr. Macdonald wrote.

> Maxfield Parrish's *Day Dreams* is replaced on the living-room wall by van Gogh's *Sunflowers* or even a Picasso print. . . . Midcult is a more dangerous opponent of High Culture because it incorporates so much of the avant-garde. The four Midcult works noticed above (i.e., those of Hemingway, Wilder, MacLeish, and Benét) were more advanced and sophisticated for their time, than were the novels of

[4] Mr. Macdonald's idiom itself needs explaining. In the early 1930's, the "tough" phase of American radicalism, a Bolshevik habit of compressing words—such as *politburo* for the political bureau of the Party, or *orgburo* for the organization bureau—caught on. Thus, the vogue of proletarian literature was known as *proletcult*. Mr. Macdonald adopted this jargon for his own sardonic style.

> John Galsworthy. They are, so to speak, the products of lapsed avant-gardists who knew how to use the modern idiom in the service of the banal. . . . Hollywood movies aren't as terrible as they once were, but they aren't as good either; the general level of taste and craftsmanship has risen but there are no more great exceptions like Griffith, von Stroheim, Chaplin, Keaton. . . .

Hannah Arendt, a thoughtful and disquieting social critic, took the classical argument one step further and blended with it a historical-Marxist analysis. She argued that bourgeois "society"—she here means the relatively homogeneous community of educated and cultivated persons—had always treated culture as a commodity and had gained snob values from its exchange, that there has always existed a certain tension between "culture" (i.e., the producers of art) and "society" (which consumed it).[5] But for her there were two crucial differences between the past and the present. In the old days, individualism flourished or was made possible through an escape *from* society, often into rebel or bohemian worlds. ("A good part of the despair of individuals under conditions of mass society is due to the fact that these avenues of escape are, of course, closed as soon as society has incorporated all the strata of the population.") Moreover, though "society" in the past coveted culture largely for its snob appeal, it did not *consume* culture, even if it abused or devaluated it and turned "cultural things into social commodities." Mass society, "on the contrary, wants not culture, but entertainment, and the wares offered by the entertainment industry are indeed consumed by society just as are any other consumer goods."

In sum, though in the 1950's there was a burning out of the radical political will, this radical will—the distancing of self from the society—was maintained in the culture and through cultural criticism. When new political impulses arose in the 1960's, radicalism found the values of the adversary culture—the attack on society through such themes as alienation—as the Ariadne's thread which allowed it to emerge into a bright, new radical era.

---

[5] Hannah Arendt, "Society and Culture," in *Culture for the Millions,* pp. 43-53. The argument is elaborated in *Between Past and Future* (New York, 1961), pp. 197-226.

III

We come to an extraordinary sociological puzzle. A single cultural temper, mood, movement—its very amorphousness or protean nature precludes a single encapsulating term—has persisted for more than a century and a quarter, nourishing renewed and sustained attacks on the social structure. Perhaps the most inclusive term for this cultural temper is *modernism:* the self-willed effort of a style and sensibility to remain in the forefront of "advancing consciousness." What is the nature, then, of this sentiment that, antedating even Marxism, has been attacking bourgeois society and has been able to sustain such a program? Why has it so captured the artistic imagination that it can preserve itself through generations, and have fresh appeal for each new cohort of the intelligentsia?

Modernism pervades all the arts. Yet if one looks at the individual examples, there seems to be no single unifying principle. It includes the new syntax of Mallarmé the dislocation of forms of Cubism, the stream of consciousness in Virginia Woolf or Joyce, the atonality of Berg. Each of these, as it first appeared, was "difficult" to understand. In fact, as a number of writers have suggested, original difficulty is a sign of a modernist movement. It is willfully opaque, works with unfamiliar forms, is self-consciously experimental, and seeks deliberately to disturb the audience—to shock it, shake it up, even to transform it as if in a religious conversion. This very difficulty is clearly one source of its appeal to initiates, for esoteric knowledge—like the special formula of the magi or the hermeticism of ancient priests—gives one an enhanced sense of power over the vulgar and unenlightened.

Modernism is a response to two social changes in the nineteenth century, one on the level of sense perception of the social environment, the other of consciousness about the self. In the everyday world of sense impressions, there was a disorientation of the sense of space and time derived from the new awareness of motion and speed, light, and sound, which came from the revolutions in communication and transport. The crisis in self-consciousness arose from the loss of religious certitude, of belief in an afterlife, in heaven or hell, and the new consciousness of an immutable boundary beyond life and the nothingness of death. In effect, these were two new ways of experiencing the world and often the artist himself was never wholly aware of the sources of disorientation in the social environment which had shaken up the world and made it seem as if there were only pieces. Yet he had to reassemble these pieces in a new way.

For the second half of the nineteenth century, then, an ordered world was a chimera. What was suddenly real, in molding the sense perception of an environment, was movement and flux. A radical change in the nature of esthetic perception had suddenly occurred. If one asks, in esthetic terms, how modern man differs from the Greeks in experiencing sensations or emotions, the answer would have to do not with the basic human feelings, such as friendship, love, fear, cruelty, and aggression, which are common to all ages, but with the temporal-spatial dislocation of motion and height. In the nineteenth century, for the first time in history, men could travel faster than on foot or on an animal, and gain a different sense of changing landscape, a succession of images, the blur of motion, which he had never before experienced. Or one could, first in a balloon and later in a plane, rise thousands of feet in the sky and see topographical patterns that the ancients had never known.

What was true of the physical world was equally true of the social. With the growth of numbers and density in the cities, there was greater interaction among persons, a syncretism of experience that provided a sudden openness to new styles of life—a geographical and social mobility—that had never been available before. In the canvases of art, the subjects were no longer the mythological creatures of the past or the stillness of nature, but the promenade and the *plage,* the bustle of city life, and the brilliance of night life in an urban environment transformed by electric light. It is this response to movement, space, and change which provided the new syntax of art and the dislocation of traditional forms.

In the classical premodern view, art was essentially contemplative; the viewer or spectator held "power" over the experience by keeping his esthetic distance from it. In modernism, the intention is to "overwhelm" the spectator so that the art product itself—through the foreshortening of perspective in painting, or the "sprung rhythm" of a Hopkins in poetry—imposes itself on the spectator in its own terms. In modernism, genre becomes an archaic conception whose distinctions are ignored in the flux of experience. In all this, there is an "eclipse of distance," so that the spectator loses control and becomes subject to the intentions of the artist. The very structural forms are organized to provide immediacy, simultaneity, envelopment of experience. Power has moved from the spectator, who could contemplate the picture, the sculpture, or the story, to the artist, who brings the viewer into his own field of action. The eclipse of distance provides a stylistic unity, a common syntax for painting, poetry, narrative, music, and becomes a common structural component—a formal element—across all the arts.

All of this was reflected in the explosive burst of artistic energy in the forty years before World War I. In the Impressionists' experiments with light, the capture of motion by the Futurists, the spatial dislocation of form in Cubism, then a bit later in the anti-art of Dadaism—in which everyday objects and "readymades" are pasted together on a canvas—one sees the bewildering succession of efforts to catch the swiftness of change through new kinds of painting. The modernist effort to capture this flux gives full meaning, I think, to Irving Howe's citation of Virginia Woolf's gnomic remark: "On or about December 1910, human nature changed." As Howe comments, in this there is a "frightening discontinuity between the traditional past and the shaken present . . . the line of history has been bent, perhaps broken."

In making this break, in the emphasis on the *absolute present*, both artist and spectator are forced to make and remake themselves anew each moment. With the repudiation of unbroken continuity, and the belief that the future is in the present, one loses the classical sense of wholeness or completeness. The fragment replaces the whole: one finds a new esthetic in the broken torso, the isolated hand, the primitive grimace, the figure cut by the frame. And in the mingling and jostling of styles, the very idea of genre and boundary, of principles appropriate to a genre, is abandoned. One might say, in fact, that esthetic disaster itself becomes an esthetic.

Enter nothingness and self

The sense of movement and change—the upheaval in the mode of confronting the world—established vivid new conventions and forms by which people judged their sense perceptions and experience. But more subtly, the awareness of change prompted a deeper crisis in the human spirit, the fear of nothingness. The decline of religion, and especially of belief in an immortal soul, provoked a momentous break with the centuries-old conception of an unbridgeable chasm between the human and the divine. Men now sought to cross that gulf and, as Faust, the first modern, put it, attain "godlike knowledge," to "prove in man the stature of a god" or else confess his "kinship with the worm."

As a consequence of this superhuman effort, in the nineteenth century, the sense of the self comes to the fore. The individual comes to be considered as unique, with singular aspirations, and life assumes a greater sanctity and preciousness. The enhancement of the single life becomes a value for its own sake. Economic meliorism, anti-slavery sentiment, women's rights, the end of child labor and cruel punishments became the social issues of the day. But in a deeper metaphysical sense, this spiritual enterprise became the basis for the idea that men could go beyond necessity, that they would no longer be constrained by nature but could arrive—in Hegel's phrase—at the end of history, in the kingdom of perfect freedom. The "unhappy consciousness" of which Hegel wrote is the realization of a divine power and status which man must strive to achieve. The deepest nature of modern man, the secret of his soul as revealed by the modern metaphysic, is that he seeks to reach out beyond himself; knowing that negativity—death—is finite, he refuses to accept it. Behind the chiliasm of modern man, is the megalomania of self-infinitization. In consequence, the modern hubris is the refusal to accept limits, the insistence on continually reaching out; and the modern world proposes a destiny that is always *beyond*—beyond morality, beyond tragedy, beyond culture.[6]

---

[6] Compare these powerful statements by two contemporary writers. In Malraux' *Man's Fate* (1933) Old Gisors describes the Baron de Clappique and his desires:

> To be more than a man in a world of men. To escape man's fate. [To be] not powerful: all powerful. The visionary disease, of which the will to power is only the intellectual justification, is the will to godhead: every man dreams of being god.

In Saul Bellow's *Mr. Sammler's Planet* (1970) old Sammler reflects:

> You wondered whether . . . the worst enemies of civilization might not prove to be its petted intellectuals who attacked it at its weakest moments—attacked it in the name of proletarian revolution, in the name of reason and in the name of irrationality, in the name of visceral depth, in the name of sex, in the name of perfect and instant freedom. For what it amounted to was limitless de-

The triumph of will

In Western consciousness there has always been tension between the rational and the nonrational, between reason and will, between reason and instinct, as the driving forces of man. A basic triadic distinction was made by Plato, who divided the soul into the rational, the spirited, and the appetitive. But whatever the specific distinctions, rational judgment was traditionally thought to be superior in the hierarchy, and this order dominated Western culture for almost two millennia.

Modernism dirempts this hierarchy. It is the triumph of the spirited, of the will. In Hobbes and Rousseau, intelligence is a slave to the passions. In Hegel, the will is the necessary component of knowing. In Nietzsche, the will is fused with the esthetic mode, in which knowledge derives most directly ("apprehended, not ascertained," as he says in the first line of *The Birth of Tragedy*) from intoxication and dream. And if the esthetic experience alone is to justify life, then morality is suspended and desire has no limit. Anything is possible in this quest of the self to explore its relation to sensibility.

The emphasis of modernism is on the present, or on the future, but never on the past. Yet when one is cut off from the past, one cannot escape the final sense of nothingness that the future then holds. Faith is no longer possible, and art, or nature, or impulse can erase the self only momentarily in the intoxication or frenzy of the Dionysian act. But intoxication always passes and there is the cold morning after, which arrives inexorably with the break of day. This inescapable eschatological anxiety leads inevitably to the feeling—the black thread of modernist thought—that each person's own life is at the end of time. The sense of an ending, the feeling that one is living in an apocalyptic age, is, as Frank Kermode has observed, "as endemic to what we call modernism as apocalyptic utopianism is to political revolution. . . . Its recurrence is a feature of our cultural tradition."[7]

---

mand—insatiability, refusal of the doomed creature (death being sure and final) to go away from this world unsatisfied. A full bill of demand and complaint was therefore presented by each individual. Non-negotiable. Recognizing no scarcity of supply in any human department.

[7] Frank Kermode, *The Sense of an Ending* (New York, 1967, p. 98.

In discussing modernism, the categories of "left" and "right" make little sense. Modernism, as Thomas Mann phrased it, cultivates "a sympathy for the abyss." Nietzsche and Yeats, Pound and Wyndham Lewis were politically far to the right. Gide was a pagan, Malraux a revolutionist. But whatever the political stripe, the modern movement has been united by rage against the social order as the first cause, and a belief in the apocalypse as the final cause. It is this trajectory which provides the permanent appeal and the permanent radicalism of that movement.

IV

Traditional modernism, in Frank Kermode's term, sought to substitute for religion or morality an esthetic justification of life; to create a work of art, to be a work of art—this alone provided meaning in man's effort to transcend himself. But in going back to art, as is evident in Nietzsche, the very search for the roots of self moves the quest of modernism from art to psychology: from the product to the producer, from the object to the psyche.

In the 1960's, a powerful current of post-modernism has developed which has carried the logic of modernism to its farthest reaches. In the theoretical writings of Norman O. Brown and Michel Foucault, in the novels of William Burroughs, Jean Genet, and to some extent Norman Mailer, and in the porno-pop culture that is now all about us, one sees a logical culmination of modernist intentions.

There are several dimensions to the post-modernist mood. Thus, against the esthetic justification for life, post-modernism has completely substituted the instinctual. Impulse and pleasure alone are real and life-affirming; all else is neurosis and death. Moreover, traditional modernism, no matter how daring, played out its impulses in the imagination, within the constraints of art. Whether demonic or murderous, the fantasies were expressed through the ordering principle of esthetic form. Art, therefore, even though subversive of society, still ranged itself on the side of order and, implicitly, of a rationality of form, if not of content. Post-modernism overflows the vessels of art. It tears down the boundaries and insists that *acting out,* rather than making distinctions, is the way to gain knowledge. The "happening" and the "environment," the "street" and the "scene," are the proper arena for life.

Extraordinarily, none of this is in itself completely new. There has always been an esoteric tradition within all Western religion which has sanctioned participation in secret rites of release, debauch, and total freedom for those—the "gnostics"—who have been initiated into secret sects through secret knowledge. Gnosticism, in its intellectual formulations, has provided the justification for the attacks on restraints that every society has imposed on its members. Yet in the past, this knowledge was kept hermetic, its members were secretive. What is most striking about post-modernism is that what was once maintained as esoteric is now proclaimed as ideology, and what was once the property of an aristocracy of the spirit is now turned into the democratic property of the mass. The gnostic mode has always beat against the historic, psychological taboos of civilization. That assault has now been made the platform of a widespread cultural movement.

The post-modern temper, looked at as a set of loosely associated doctrines, itself goes in two directions. One is philosophical, a kind of negative Hegelianism. Michel Foucault, who is now very much "in," sees man as a short-lived historical incarnation, "a trace on the sand," to be washed away by the waves. The "ruined and pest-ridden cities of man called 'soul' and 'being' will be deconstructed." It is no longer the decline of the West, but the end of all civilization. Much of this is modish, a play of words pushing a thought to an absurd logicality. Like the angry playfulness of Dada or Surrealism, it will probably be remembered, if at all, as a footnote to cultural history.

But the post-modern temper, moving in another direction, does carry a much more significant implication. It provides the doctrinal spearhead for an onslaught on the values and motivational patterns of "ordinary" behavior, in the name of liberation, eroticism, freedom of impulse, and the like. It is this, dressed up in more popular form, which is the real importance of the post-modernist doctrine. For it means that a crisis of middle-class values is at hand.

Death of the bourgeois

The bourgeois world-view—rationalistic, matter-of-fact, pragmatic; neither magical, mystical, nor romantic; emphasizing work and function; concerned with restraint and order in morals and conduct—had by the mid-nineteenth-century come to dominate, not only the social structure (the organization of the economy), but also the culture, especially the religious order and the educational system which instilled "appropriate" motivation in the child. It reigned triumphant everywhere, opposed only in the realm of culture by those who disdained its un-heroic and anti-tragic mood, as well as its orderly attitude towards time.

The last hundred years has seen an effort by anti-bourgeois culture to achieve *autonomy* from the social structure, first by a denial of bourgeois values in the realm of art, and second by carving out enclaves where the bohemian and the *avant-gardist* could live a contrary style of life. By the turn of the century the *avant-garde* had succeeded in establishing a "life-space" of its own, and by 1910-1930 it was on the offensive against traditional culture.

Today, in both doctrine and life-style, the anti-bourgeois has won. This triumph means that, in the culture today, antinomianism and anti-institutionalism rule. In the realm of art, on the level of esthetic doctrine, no one opposes the idea of boundless experiment, of unfettered freedom, of unconstrained sensibility, of impulse being superior to order, of the imagination being immune to merely rational criticism. There is no longer an *avant-garde,* because no one in our post-modern culture is on the side of order or tradition. There exists only a desire for the new.

The traditional bourgeois organization of life—its rationalism and sobriety—no longer has any defenders in the culture, nor does it have any established system of culture meanings or stylistic forms with any intellectual or cultural respectability. To assume, as some social critics do, that the technocratic mentality dominates the cultural order is to fly in the face of every bit of evidence at hand. What we have today is a radical disjunction of culture and social structure, and it is such disjunctions which historically have paved the way for more direct social revolutions.

In two fundamental ways, that revolution has already begun. First, the autonomy of culture, achieved in art, now passes over into the arena of life. The post-modernist temper demands that what was previously played out in fantasy and imagination must be acted out in life as well. There is no distinction between art and life. Anything permitted in art is permitted in life as well.

Second, the life-style once practiced by a small *céna-*

cle, whether the cool life-mask of a Baudelaire or the hallucinatory rage of a Rimbaud, is now copied by a "many"—a minority in the society to be sure, but nonetheless large in number—and dominates the cultural scene. This change of scale gives the culture of the 1960's its special power, plus the fact that a bohemian life-style once limited to a tiny elite is now acted out on the giant screen of the mass media. Woodstock—both the event and the movie—gives us a clear sense of what's happening.

The combination of these two changes adds up to the beginning of a major onslaught by the "culture" against the "social structure." When such attacks were launched before—say, André Breton's surrealistic proposal in the early 1930's that the Towers of Notre Dame be replaced by an enormous glass cruet, one of the bottles filled with blood, the other with sperm, the church itself becoming a sexual school for virgins—they were understood as heavy-handed japes, perpetrated by the licensed "fools" of society. But the rise of a hip-drug-rock culture on a popular level, and the "new sensibility" of black-mass ritual and violence in the arena of culture, are a set of cultural actions that undermine the social structure itself by striking at the motivational and psychic-reward system which has sustained it. In this sense, the culture of the 1960's has a new and perhaps distinctive historic meaning.

V

Changes in cultural *ideas* have an immanence and autonomy because they develop from an internal logic at work with a cultural tradition. In this sense, new ideas and forms derive out of a kind of dialogue with, or rebellion against, previous ideas and forms. But changes in cultural *practice and life-styles* necessarily interact with social structure, since works of art, accoutrements, records, films, and plays are bought and sold in the market. The market is where social structure and culture cross. Changes in culture—particularly the emergence of new life-styles—are made possible, not only by changes in sensibility, but by shifts in the social structure itself. One can see this most readily, in the contemporary instance, in the development of new buying habits in a high consumption economy and the resultant erosion of the Protestant Ethic and the Puritan Temper, the two codes which once sustained the traditional value system of our society. It is the breakup of this ethic and temper, owing as much to changes in social structure as in the culture, that has undercut the beliefs and legitimations that sanctioned work and reward in American society. It is this transformation and the lack of any rooted new ethic, that is responsible, in good part, for the sense of disorientation and dismay that marks the public mood today.

The "Protestant Ethic" and the "Puritan Temper" were codes that emphasized work, sobriety, frugality, sexual restraint, and a forbidding attitude toward life. They defined the nature of moral conduct and social respectability. The post-modernist culture of the 1960's has been interpreted, because it calls itself a "counter-culture," as defying the Protestant Ethic, heralding the end of Puritanism, and mounting a final attack on bourgeois values. This is too facile. The Protestant Ethic and the Puritan Temper, as social facts, were eroded long ago, and they linger on as pale ideologies, used more by moralists to admonish and by sociologists to mythologize than as behavioral realities. The breakup of the traditional bourgeois value system, in fact, was brought about by the bourgeois economic system—by the free market, to be precise.

From the Protestant ethic to the psychedelic bazaar

The Protestant Ethic and the Puritan Temper in the United States were the world view of an agrarian, small-town, mercantile, and artisan way of life. In the United States, as Page Smith reminds us, "if we except the family and the church, the basic form of social organization up to the early decades of the twentieth century was the small town."[8] The life and character of American society were shaped by the small town—and especially by its religions. The erosion of traditional (i.e., smalltown) American values took place on two levels. In the realm of culture and ideas, a withering attack on small-town life as constricting and banal was first organized, in the period between 1910 and 1920, by the Young Intellectuals, a self-consciously defined group, including such figures as Van Wyck Brooks and Harold Stearns, who sought a new and more inclusive vision of American culture. This attack was sustained in the journalistic criticism of H. L. Menck-

[8] Page Smith, *As a City Upon a Hill* (New York, 1966), pp. vii-viii.

en and in the sketches and novels of Sherwood Anderson and Sinclair Lewis.

But a more fundamental transformation was occurring in the social structure itself. There was, first, the enormous expansion of the cities in response to industrialism. Equally important, if not more so, was the change in the motivations and rewards of the system itself. The rising wealth of the plutocracy, becoming evident in the Gilded Age, meant that work and accumulation were no longer ends in themselves (though they were still crucial to a John D. Rockefeller or an Andrew Carnegie) but means to consumption and display. Status and its badges, not work and the election of God, became the mark of success.

This is a familiar process of social history, with the rise of new classes, though in the past it was military predators whose scions went from spartan to sybaritic living. Because the parvenu classes could distance themselves from the rest of society, such social changes often developed independently of changes in the lives of the classes below. But the real social revolution in modern society came in the 1920's, when the rise of mass production and high consumption began to transform the life of the middle class itself. In effect the Protestant Ethic as a social reality and a life-style for the middle class was replaced by a materialistic hedonism, and the Puritan Temper by a psychological eudaemonism.

But bourgeois society, justified and propelled as it had been in its earliest energies by these older ethics, could not easily admit to the change. It promoted a hedonistic way of life furiously—one has only to look at the transformation of advertising in the 1920's—but could not justify it. It lacked a new religion or a value system to replace the old, and the result was a disjunction.

The "new capitalism"—the phrase was used in the 1920's—continued to demand a Protestant Ethic in the area of production—that is, in the realm of work—but to stimulate a demand for pleasure and play in the area of consumption. The disjunction was bound to widen. The spread of urban life, with its variety of distractions and multiple stimuli; the new roles of women, created by the expansion of office jobs and the freer social and sexual contacts; the rise of a national culture through motion pictures and radio—all contributed to a loss of social authority on the part of the older value system.

The Puritan Ethic might be described most simply by the phrase "delayed gratification," and by restraint in gratification. It is, of course, the Malthusian injunction for prudence in a world of scarcity. But the claim of the American economic system was that it had introduced abundance, and the nature of abundance is to encourage prodigality rather than prudence. The "higher standard of living," not work as an end in itself, then becomes the engine of change. The glorification of plenty, rather than the bending to niggardly nature, becomes the justification of the system. But all of this was highly incongruent with the theological and sociological foundations of nineteenth century Protestantism, which was in turn the foundation of the American value system.

The abdication of the corporate class

The ultimate support for any social system is the acceptance by the population of a moral justification of authority. The older justifications of bourgeois society lay in the defense of private property, which itself was justified on the grounds, elaborated by Locke, that one infused one's own labor into property. But the "new capitalism" of the twentieth century has lacked such moral grounding, and in periods of crisis it has either fallen back on the traditional value assertions, which have been increasingly incongruent with social reality, or it has been ideologically impotent.

It is in this context that one can see the weakness of corporate capitalism in trying to deal with some of the major political dilemmas of the century. Political—and value—conflicts in the United States can be looked at from two different perspectives. From one, there have been economic and class issues which divided farmer and banker, worker and employer, and led to the functional and interest-group conflicts which were especially sharp in the 1930's. Along a different sociological axis, one can see the politics of the 1920's, and to some extent that of the 1950's within the framework of "tradition" versus "modernity," with the rural, small-town Protestant intent on defending his historic values against the cosmopolitan liberal interested in reform and social welfare. The issues here are not primarily economic but sociocultural. The traditionalist defends fundamentalist religion, censorship, stricter divorce, and anti-abortion laws; the modernist is for secular rationality, freer personal relations, tolerance of sexual deviance, and the like. These represent the political side of cultural issues, and to the extent that culture is the symbolic expression and justification of experience, this is the realm of symbolic or expressive politics.

In this respect, the great symbolic issue of American politics was Prohibition. It was the major—and almost the last—effort by small-town and traditionalist forces to impose a specific value, the prohibition of liquor, on the rest of the society; and, initially, of course, the traditionalists won. In a somewhat different sense, McCarthyism in the 1950's represented an effort by some traditionalist forces to impose a uniform political morality on the society through conformity to one ideology of Americanism and a virulent form of anti-Communism.

Now, the curious fact is that the "new capitalism" of abundance, which emerged in the 1920's, has never been able to define its view of these cultural-political issues, as it had of the economic-political conflicts. Given its split character, it could not do so. Its values derive from the traditionalist past, and its language is the archaism of the Protestant Ethic. Its technology and dynamism, however, derive from the spirit of modernity—the spirit of perpetual innovation, and of the creation of new "needs" on the installment plan. The one thing that would utterly destroy the "new capitalism" is the serious practice of "deferred gratification."

When members of the corporate class have taken a stand on cultural-political issues, they have often divided on geographical lines. Midwesterners, or Texans, or those coming from small-town backgrounds, display traditionalist attitudes; Easterners, or products of Ivy League schools, are more liberal. More recently, the division has been based on education and age rather than region. But the singular fact remains. The new capitalism was primarily responsible for transforming the society, and in the process undermined the Puritan Temper, but it was never able to develop successfully a new ideology congruent with the change, and it used—and often was trapped by—the older language of Protestant values.

The forces of modernity, which had taken the lead against the traditionalists on these social and cultural issues, have been a *mélange* of intellectuals, professors, welfare and reform-minded individuals (though, paradoxically, the prohibition movement at its inception was allied with the reformers against the evils of industrialism and city life), joined, for political reasons, by labor leaders and ethnic politicians who represented urban forces. The dominant philosophy has been liberalism, which included a critique of the inequalities and social costs generated by capitalism.

The fact that the corporate economy has no unified value system of its own, or still mouthed a flaccid version of Protestant virtues, meant that liberalism could go ideologically unchallenged. In the realm of culture, and of cultural-social issues—of political philosophy, in short—the corporate class had abdicated. The important consideration is that, *as an ideology,* liberalism had become dominant over these past decades.

VI

From a *cultural* point of view, the politics of the 1920's to 1960's was a struggle between tradition and modernity. In the 1960's a new cultural style appeared. Call it psychedelic or call it, as its own protagonists have, a "counter-culture." It announced a strident opposition to bourgeois values and to the traditional codes of American life. "The bourgeoisie," we are told, "is obsessed by greed; its sex life is insipid and prudish; its family patterns are debased; its slavish conformities of dress and grooming are degrading; its mercenary routinization of life is intolerable. . . ."[9]

What is quixotic about such pronouncements is the polemical and ideological caricature of a set of codes that had been trampled on long ago—beginning sixty years earlier, with the Young Intellectuals. Yet such a caricature is necessary to make the new counter-culture seem more daring and revolutionary than it is. The new sensibility, with its emphasis on psychedelic experience, sexual freedom, apocalyptic moods and the like, thinks of itself as being against "bourgeois" culture. But in truth, bourgeois culture vanished long ago. What the counter-culture embodies is an extension of the tendencies initiated sixty years ago by political liberalism and modernist culture, and represents, in effect, a split in the camp of modernism. For it now seeks to take the preachments of personal freedom, extreme experience ("kicks," and "the high")

---

[9] Theodore Roszak, *The Making of a Counter-Culture,* (Doubleday, 1969) p. 13.

and sexual experimentation, to a point in *life-style* that the liberal culture—which would approve of such ideas in *art and imagination*—is not prepared to go. Yet liberalism finds itself uneasy to say why. It approves a basic permissiveness, but cannot with any certainty define the bounds. And this is its dilemma. In culture, as well as in politics, liberalism is now up against the wall.

Liberalism also finds itself in disarray in an arena where it had joined in support of capitalism—in the economy. The economic philosophy of American liberalism had been rooted in the idea of growth. One forgets that in the late 1940's and 1950's Walter Reuther, Leon Keyserling, and other liberals had attacked the steel companies and much of American industry for being unwilling to expand capacity and had urged the government to set target growth figures. Cartelization, monopoly, and the restriction of production had been historic tendencies of capitalism. The Eisenhower administration consciously chose price stability over growth. It was the liberal economists who instilled in the society the policy of the conscious planning of growth through government inducements (e.g., investment credits, which industry, at first, did not want) and government investment. The idea of potential GNP and the concept of "short-fall"—the posting of a mark of what the economy at full utilization of resources could achieve compared to the actual figure—was introduced in the Council of Economic Advisors by the liberals. The idea of growth has become so fully absorbed as an economic ideology that one realizes no longer, as I said, how much of a liberal innovation it was.

The liberal answer to social problems such as poverty was that growth would provide the resources to raise the incomes of the poor.[10] The thesis that growth was necessary to finance public services was the center of John Kenneth Galbraith's book *The Affluent Society*.

And yet, paradoxically, it is the very idea of economic growth that is now coming under attack—and from liber-

---

[10] More technically, it was based on the welfare economics theorem of Pareto optimality, namely that one should seek a condition where some people would be better off without anyone being worse off. The direct redistribution of income is politically difficult if not impossible. However, from new or added national income, a higher proportion of the gains can be used to finance social welfare programs; and this was, as Otto Eckstein pointed out in "The Economics of the Sixties," *The Public Interest*, No. 19, Spring 1970, precisely what Congress was willing to do when economic growth was resumed in the Kennedy administration.

als. Affluence is no longer seen as an answer. Growth is held responsible for the spoliation of the environment, the voracious use of natural resources, the crowding in the recreation areas, the densities in the city, and the like. One finds, startlingly, the idea of zero economic growth—or John Stuart Mill's idea of the "stationary state"—now proposed as a serious goal of government policy. Just as the counter-culture rejects the traditional problem-solving pragmatism of American politics, it now also rejects the newer, liberal policy of economic growth as a positive goal for the society. But without a commitment to economic growth, what is the *raison d'être* of capitalism?

Two crises

American society faces a number of crises. Some are more manifest—the alienation of the young, the militancy of the blacks, the crisis of confidence created by the Vietnam war. Some are structural—the creation of a national society, a communal society, and a post-industrial phase—which are reworking the occupational structure and the social arrangements of the society.[11] These are all aspects of a political torment in the social system. Yet these crises, I believe, are manageable (not solvable; what problems are?) if the political leadership is intelligent and determined. The resources are present (or will be, once the Vietnam war is ended) to relieve many of the obvious tensions and to finance the public needs of the society. The great need here is *time*, for the social changes which are required (a decent welfare and income maintenance system for the poor, the reorganization of the universities, the control of the environment) can only be handled within the space of a decade or more. It is the demand for "instant solutions" which, in this respect, is the source of political trouble.

But the deeper and more lasting crisis is the cultural one. Changes in moral temper and culture—the fusion of imagination and life-styles—are not amenable to "social engineering" or political control. They derive from the value and moral traditions of the society, and these cannot be "designed" by precept. The ultimate sources are the

---

[11] For a detailed discussion of some of the "structural revolutions" which underlie the more manifest crises, and for a discussion of political dilemmas of liberalism, see my essay "Unstable America" in the June 1970 issue of *Encounter* (London).

religious conceptions which undergird a society; the proximate sources are the "reward systems" and "motivations" (and their legitimacy) which derive from the arena of work (the social structure).

American capitalism, as I have sought to show, has lost its traditional legitimacy which was based on a moral system of reward, rooted in a Protestant sanctification of work. It has substituted in its place a hedonism which promises a material ease and luxury, yet shies away from all the historic implications which a "voluptuary system"—and all its social permissiveness and libertinism—implies.

This is joined to a more pervasive problem derived from the nature of industrial society. The characteristic style of an industrial society is based on the principles of economics and economizing: on efficiency, least cost, maximization, optimization, and functional rationality. Yet it is at this point that it comes into sharpest conflict with the cultural trends of the day, for the culture emphasizes anticognitive and anti-intellectual currents which are rooted in a return to instinctual modes. The one emphasizes functional rationality, technocratic decision-making, and meritocratic rewards. The other, apocalyptic moods and antirational modes of behavior. It is this disjunction which is the historical crisis of Western society. This cultural contradiction, in the long run, is the deepest challenge to the society.

3

The new markets and the new capitalism

PETER F. DRUCKER

THE third merger wave to wash over the American economy in this century is now receding fast. It leaves behind a landscape changed even more fundamentally in its economic structure than its predecessors did, sixty-five and forty years ago respectively.

The first merger wave, the one that reached its climax between 1900 and 1910, was that of the "tycoons," with J. P. Morgan's US Steel and John D. Rockefeller's Standard Oil Company as the prototypes. In these mergers a dominant industrialist or financier tried to occupy a commanding height in the economy by obtaining control of a major material or a major industry. They were "offensive" mergers.

The second merger wave, in the 1920's was, by contrast, "defensive" in character. Its prototype, indeed its earliest example, was General Motors, put together between 1910 and 1920 through the merger of medium-sized car companies for common defense against Henry Ford's near-monopoly. The aim of these "defensive" mergers was to create in various industries a "number two" which would be able to hold its own against the giants which the first merger wave had spawned. And while a good many of the creations of this period—such as General Motors itself—in turn became the leading company in their industry, the "defensive" mergers made for less concentration of power in the country's major industries and for more vigor and equality in competition. It often resulted in "oligopoly"; but it more often thwarted "monopoly."

The third merger wave, the one which has now peaked, began with "defensive" mergers, very similar to those of the 1920's. Typical are the railroad mergers, such as the one which created the Penn-Central and the new railroad system in the Northwest between Chicago and Seattle. (It is no coincidence that the plans for these railroad mergers are 40 years old—that is, they go back to the period of "defensive" mergers in the 1920's.) The mergers among the major New York banks in the late 1940's and early 1950's—such as the merger that put the Chase Bank and the Bank of Manhattan together into the Chase Manhattan Bank, the National City Bank and the First National into the First National City, or the Chemical Bank and the New York Trust Company into Chemical New York—were also "defensive." They have resulted in commercial banking in New York—and commercial banking internationally—becoming far more competitive. There are fewer players, to be sure, but they are far stronger and far more aggressive. These "defensive" mergers of the last 20 years aimed, as did the mergers of the 1920's, at creating enterprises large enough for a national or international market and strong enough to withstand competition on a national (or, as in the case of the banks, an international) scale. Mergers of this kind have been taking place also in manufacturing industry. The merger, for instance, between two medium-sized forest products companies, Champion Paper and US Plywood, created a billion dollar business producing all kinds of forest products from saw timber and plywood veneer to fine paper. Still, the combined company is considerably smaller than the country's largest paper company, International Paper, itself created by mergers before World War II.

The "diversification merger"

But the more typical mergers of the post-war period, and especially of the last ten years, resemble neither the "offensive" merger of 1900 nor the "defensive" merger of 1925.

To begin with, there were two very different and yet "typical" kinds. In the "diversification merger" which dominated the early years of the period—say, through 1964 or so—a large company, though distinctly not one of the leaders in its industry, merges with companies of similar size but in totally different lines of business. One leader in this kind of merger activity has been IT&T—the initials standing for the original International Telegraph and Telephone. Originally this was a company operating telephone businesses abroad, especially in the Latin-speaking countries. It also manufactured telephone equipment for these companies. It was, in fact, originally the foreign counterpart to the American Telephone and Telegraph Company, the Bell System. During the last 40 years, the company gradually gave up or lost one operating telephone company after the other. It became instead a world-wide manufacturer of electronics, very large in its total output, but no better than third or fourth in every one of its markets or technologies. And then, in the last 10 or 15 years, it expanded through the acquisition of companies that have nothing to do, at least at first glance, with its main business. It acquired Avis-Rent-A-Car, the Sheraton hotel chain, Levitt—the country's largest mass-builder. It plans to merge with Hartford Fire Insurance, one of the large casualty insurance companies (a merger blocked so far by the Antitrust division).

In a similar "diversification through merger," American Radiator & Standard Sanitary Corporation, an old and large, but stagnant, producer of plumbing and heating equipment merged, first with Westinghouse Air Brake, a large and also rather stagnant producer of railroad brakes and signals, then with Mosler Safe, a medium-sized and rapidly growing company making mostly equipment for banks, and then with Lyons, a California-based mass-builder. In the process the company became "American Standard."

Two other of the many hundreds of "diversification mergers" further exemplify the phenomenon. One was the merger in which Montgomery Ward, a large retail and mail-order chain, came together with Container Corporation of America to form a new company, Marcor. The other example is the acquisition of the country's second-largest medium-term finance company, Commercial Credit, by Control Data, the one computer manufacturer other than IBM (though very much smaller) which has so far shown any ability to make profits out of making computers. At first sight, Montgomery Ward and Container Corporation have nothing in common, and their merger makes no economic sense. But both companies, while absolutely very large, are no more than "also-rans" in their respective industries, with less than one-third of the sales of the industry leader (Sears Roebuck and American Can, respectively). In their problems, their opportunities, and their strategic decisions, they may therefore be akin to each other. Similarly, producing big computers and providing installment loans to automobile buyers have, it seems, nothing in common. But the central problem of the successful computer manufacturer is to finance his machines which, as a rule, are leased rather than sold. The central problem of a finance company, especially the smaller and weaker one, is a dependable and steady supply of high-quality borrowers. Again, there is high degree of "fit" even though the respective businesses are as diverse as can be.

The "takeover merger"

"Diversification mergers," though continuing right through the 1960's passed their peak around 1964 or 1965. From then on, until the end of 1969, the mergers that made the headlines were something quite different, "merger by takeover." This merger is forced upon a reluctant, and often loudly resisting, management by organizing a stockholders' revolt against it. And the one who "takes over" is almost invariably a very much smaller company, a total outsider—indeed, typically a brash newcomer who did not even exist a few years earlier. In the "diversification mergers," both parties plighted their troth with the promise of "synergism" which would somehow make the combined business more productive than the two had been alone. But in the "takeover," the justifying slogan is "asset management," that is, the maximization of the value of the shareholders' equity through financial management. In effect, the "takeover" is far less a merger of businesses than a *coup d'état*. A guerrila leader, himself owning practically no part of the company he acquires, gets the outside shareholders of large publicly-owned companies to oust their own "professional business management" and put him into the saddle. The more successful of these new corporate guerrilla captains went from one "takeover" to another. Starting with nothing, they built "conglomerates" showing total revenues in the billions.

Among the "takeover" victims have been some of the oldest and best-known companies in the country, companies run by entrenched "professional management." They include two of the world's largest steel companies, Jones & Laughlin and Youngstown Sheet & Tube, both with sales around the billion dollar mark. They were taken over respectively by Ling-Temco-Vought, built by James Ling in a few years from a small electronics shop into a medium-sized aerospace company with sales around $160 million, and by Lykes Brothers Steamship, a New Orleans-based shipping line which never in all its 40 years of history had had sales of more than $70 million.

In another bitterly fought takeover, AMK, a company of whom a few years earlier nobody had even heard, took over the old United Fruit Company in Boston—a company with over $400 million in assets and almost $400 million in sales. (This was AMK's second leap—in the first one a year earlier, it had taken over one of the oldest meat-packers, Morrell.) AMK's base had been a small company making industrial machinery. The United Fruit shareholders sided with the raider, not because their management had failed, but because it had been so successful in its attempt to turn around and save an old and ailing enterprise that it had accumulated a large amount of cash. The most spectacular of these takeovers would have been (it never came off) the takeover of the country's sixth-largest bank, Chemical New York—with assets of $9 billion—by a company, called Leasco, that was not even mentioned in the financial handbooks of 1966. It had started out, a few years earlier, as a small computer leasing operation without any capital to speak of and with fewer than a dozen employees.

There thus came into being a whole new group of entrepreneurs. They are not "owners," but they know how to mobilize the vast multitude of shareholders of big publicly-owned companies against management. They have been able again and again to unseat "professional management" in the name of "asset management," that is, by promising to maximize financial returns. And their aim is not the "synergism" (whatever that may mean) of "diversification." It is the "conglomerate" built by financial manipulation and based on financial control.

The new "growth" companies

Perhaps even more significant, however, is another development, and one that never before coincided with a wave of mergers. It is the emergence of yet another group of new entrepreneurs, far more numerous than the "asset managers" and perhaps a good deal sounder, though neither as colorful nor as spectacular. These are the men who have been building new "growth" businesses in very large numbers. New businesses are, of course, being started all the time. But these new businesses have been started as "growth" businesses, and from the beginning were supported by large investments from the capital market. In every year from 1965 to 1969, 8-10,000 brand-new "growth" businesses got going. These entrepreneurs went to the capital market for anything up to $1 million before their business had even been started, produced its first product, or made its first sale. A year or two later most came back for another substantial sum of money, ranging from $1 million to $10 million apiece. These companies were still sufficiently small and their investors sufficiently few in number—and were also what the securities laws call "sophisticated investors" (that is primarily investment institutions)—not to have to register their securities with the Securities Exchange Commission. Yet they were sufficiently large already to have to apprise the Commission of their existence. All told, these new businesses raised about $5 billion to $10 billion annually from the "sophisticated investors" during the last five years.

"Science-based" companies, most nonfinancial readers will instantly think. And indeed the "science-based" companies that sprang up in the 1950's around Boston on Route 128 or on the Peninsula south of San Francisco were the forerunners. But though "science-based" industries (such as "learning" or computer application) are to be found among the new "growth" ventures of the 1960's, they constitute a small fraction of the total. Among the "glamour" stocks for which the "sophisticated investors" bid were franchise restaurants, magazine and book publishers, nursing homes and hospitals, prefabricated housing and mobile homes, manufacturing, and others.

Some of these new "growth" companies are even to be found in finance, both on Wall Street and as managers of investment trusts. The first of these financial growth ventures—and the most successful one to date—Donaldson, Lufkin & Jenrette, was started ten years ago by a group of young business school graduates. It had become by 1969 the seventh largest Stock Exchange firm. Then it singlehandedly achieved what Franklin D. Roosevelt, with all the power of the United States government behind him, had failed to bring about 30 years earlier: to force the Stock Exchange out of being a "private club." When Donaldson, Lufkin & Jenrette outgrew its capital base in 1969 and threatened to quit the Exchange unless permitted to sell shares to the public—something always strictly forbidden by the rules which, in effect, limited investment in Stock Exchange firms to wealthy individuals—it had become so important that the Stock Exchange had to give way. Donaldson, Lufkin & Jenrette raised $12 million by selling shares to the public last April.

Very few of these new companies can be compared with Xerox, the growth company *par excellence* of the American economy in post-World War II—a company having barely $1 million in sales as recently as 1950, still having less than $15 million in sales in 1960, and, in 1969, reporting sales of $1½ billion. But a good many of these new companies grew, within a short period, to respectable middle size—$50 million, $60 million, $70 million, sometimes even $100 million in sales. An even larger number grew to the point where their founders could sell them at a considerable capital gain to older, staid and less "dynamic" companies bent on "diversification," or could, in a few cases, become "takeover entrepreneurs" in their own right. Not since the railroad and banking ventures of the Age of Jackson has there been any comparable explosion of new ventures getting, from the start, broad financial support in large amounts.

What "could not have happened"

Neither the "takeover merger" nor the new "growth" ventures could really have happened, according to "what everybody knows" about the structure of the American economy. This is brought out clearly by the cleavage between the actual developments and the magisterial announcements regarding what could happen—just when the actual developments were approaching their peak.

Nineteen sixty-seven was the year in which takeovers exploded and in which also the largest number of "new growth" businesses appeared on the capital market. It was also the year which produced the all-time best-seller by an American academic economist, John Kenneth Galbraith's *The New Industrial State*. This book has two fundamental theses. One, professional management in the big corporation is so firmly entrenched that it cannot be challenged, let alone be overthrown, from inside or outside. The dispersed "public" stockholder is completely disenfranchised, to the point where management need not, and indeed does not, aim at maximizing profitability, but can run the business comfortably to perpetuate itself in power. Secondly, *The New Industrial State* asserted that new businesses simply cannot come into existence in this economy of large corporations which manipulate the market, both that of goods and of capital. And such small new businesses as did manage to get born certainly could not possibly grow.

What makes the contrast between the theses of this best-selling book and the reality of the very moment when it appeared particularly significant is, however, that Galbraith in this book was not the innovator and iconoclast, and the exploder of the "conventional wisdom," which he had been in his earlier books. The two theses of *The New Industrial State*, however provocatively phrased by Galbraith, were the most conventional and most widely accepted theses regarding American economic structure. They go back indeed to the years before World War I, when John R. Commons, the father of American institutional economics, first propounded them. They underlay, of course, Veblen's work in the years of World War I. They were given full documentation in the classic on American corporate structure, Berle & Means' *The Modern Corporation and Private Property*, which came out in 1932. They were restated in the three books which initiated, one way or another, the tremendous interest in and study of the American business corporation in the last 25 years; James Burnham's *The Managerial Revolution* (1941) and my own books, *The Concept of the Corporation* (1946), and *The New Society* (1950). For once, in other words, Galbraith in *The New Industrial State* was the very voice of the "conventional wisdom." But this makes it all the more apparent that something significant must have happened in the very structure of the American economy in these last few years.

Indeed, not even the "diversification merger" is truly compatible with the prevailing and generally accepted doctrine of "managerialism." The doctrine preaches that entrenched management in the large publicly-owned company does not need the outside financial market. Large, well-established companies, so said the received wisdom of the last 40 years, are capable of financing themselves through "retained earnings." But if this was indeed correct, the "diversification mergers" would not, indeed could not, have happened. The accepted doctrine also preaches that such a company does not have to compete for management. Management can perpetuate itself and can offer competent mediocrities safe careers. Yet in these mergers, one top management voluntarily abdicated—for of course the new, merged company needed only one top management. No top management would commit suicide if immune to stockholder control in the first place, and capable of providing itself with the resources for its own security of tenure. And indeed none has ever done so, under these conditions. It's just that these conditions are not—or are no longer—those of the real world.

Unlike the "defensive" mergers of yesterday, the "diversification mergers" do not strengthen a company in the markets for its products or in manufacturing efficiency. The only explanation why managements in companies that are apparently doing quite well—as in the case of Commercial Credit and Control Data, Montgomery Ward and Container Corporation, Sheraton Hotels, American Radiator and Westinghouse Air Brake—might be willing and often eager to merge, is that they find themselves under pressures they cannot neglect. They are unable to attract resources they must have to survive and which they cannot generate just by being big and established. The "diversification merger," like the "takeover" and the new "growth entrepreneurship," was a seismic disturbance that argues some major structural shift someplace deep below the economy's surface.

The multinational company

One more, equally significant, development occurred in economic structure during these last few years—and it too "could not have happened." Fifteen years ago all but a handful of the major American businesses were entirely "American" (or at least "North American," i.e., with a subsidiary in Canada) in their geographic distribution. Today the great majority of major manufacturing companies are "multinational," with 20 per cent to 50 per cent of their output produced outside of the United States. Indeed, as Jean-Jacques Servan-Schreiber pointed out in his book, *The American Challenge* (which followed Galbraith's *The New Industrial State* as the international economics best-seller for 1968), the American companies producing in Europe are the world's third largest industrial power, out-produced only by the United States and Russia, and in turn out-producing even Japan and Germany. Nor is "multinationalism" confined to manufacturing companies. Large American banks—the Bank of America, the First National City Bank, and the Chase Manhattan—have today an even larger proportion of their business outside the United States than have most multinational manufacturing companies. Several Stock Exchange houses are also—White Weld, for instance—truly "multinational" and have become leading underwriters in the European capital markets. And then there are the "off shore" investment trusts, American-managed but domiciled outside the country and confined by law to doing business exclusively with non-Americans. One of them, Investors Overseas Service, started only in 1956 by a former social worker from Philadelphia, Bernard Cornfeld, had, by the end of 1969, amassed almost $2½ billion in assets and had become the leading asset-manager in many European and Latin American countries.

This development began in the United States; and for the first few years "multinational" was synonymous with "US-based." Around 1965, however, the move to "multinationalism" became truly "multinational." The fastest growers these last few years have been the Swedes. Today, three out of every ten men working for Swedish-owned manufacturing plants work outside of Sweden—a few short years ago, the figure was one out of ten! Then the Japanese, around 1968, began to move. Every issue of the *Oriental Economist*, Japan's counterpart of the *Wall Street Journal*, reports a new manufacturing plant, built by a Japanese company abroad, a new joint-venture with a non-Japanese company to produce abroad, a new Japanese manufacturing subsidiary abroad. The development has been so sweeping that one student, Judd Polk, an economist at the International Chamber of Commerce, argues that we should replace the old theory of international trade, which deals with the "international distribution of the fruits of production," with a new theory of the international allocation of the factors of production. Another observer, Howard Perlmutter of the University of Pennsylvania, predicts that, by 1985 or so, the bulk of the world's supply of manufactured goods will come from 300 world-wide companies, producing in all the major countries, managed multinationally, and owned by shareholders from all the major countries.

More than half of the production of the US-based multinationals outside of the United States has been added in the last five years. Yet five years ago, every economist "knew" that American multinational expansion had been stopped, that indeed the American multinational company was in for a sharp contraction. For in 1965 the United States government banned further investment of United States funds in American multinational businesses abroad, especially in the developed countries (other than Canada and Japan). The ostensible reason was concern for the balance of payments—though a major reason was surely also the desire to placate our European allies, especially de Gaulle, who were complaining loudly that we were buying up too many European economies. The ban has been strictly policed and is faithfully observed. Yet every year since 1965, the multinationals have invested more in Europe. The explanation is simply that America never financed the European acquisitions and new business abroad of the "multinationals;" the Europeans did it all along. They exchanged their holdings in their own national companies, trading in a restricted national market, for holdings in a multinational company that was world-wide in scope and management. During these last ten years Europe actually invested quite a bit more—a billion or two—in shares of US-based companies than US-based companies invested in production in Europe (whether through starting a European business or through acquisition of an existing one). The result of this is that Europeans, in the aggregate, now probably own as much of leading American businesses—up to 20 per cent in some cases—as American businesses own of major industries in Europe. So far, no one has paid much attention to this. But predictably we will one day soon discover it; and then "foreign domination of American industry" is likely to become as much of a political slogan in this country as "American domination of French (German, British, Italian, Dutch, etc.) industry" has become a political slogan abroad.

The Swedish and Japanese examples are even more amazing. In both countries the government has all along exercised the tightest control over investment abroad. In both countries investment in manufacturing plants abroad was officially banned and currency for it was simply not available. The "multinationalism" of industry in these two countries is, in other words, also being financed by the "multinational" investors—especially the investor in the countries in which investment is being made—who exchanges his ownership of a local business against a share in the ownership of a multinational one.

But this is as incompatible with the "verities" of international economic theory as the "takeover" wave or the emergence of the new entrepreneurs are incompatible with the received wisdom of the doctrine of "managerialism." It is not only Keynesian theory that assumes the fiscal and financial sovereignty of the national state—all economic theory these last 50 years has done that. But no sooner did the most powerful state, the United States, exercise this sovereignty, than—totally unplanned—the "Euro-dollar market" sprang up through which Europeans channeled their capital into American-based multinational companies, thus defeating the intentions of both the United States and their own governments as well.

It can be argued that the new mergers, whether "diversification" or "takeovers," are "temporary phenomena," and also that many of these were neither sound nor desirable. It can be argued that the new entrepreneurs were simply the froth on a stock market boom that is now over. It can be argued—as General de Gaulle did—that the "multinational" company is an abomination and a flagrant violation of the immutable laws of politics and history. These arguments are not without substance. There is indeed little doubt that a good many of the conglomerates were jerry-built, the result of financial sleight of hand and "asset exploitation" rather than "asset management." A good many of the new businesses, the shares of which were eagerly bought up by "sophisticated investors," were certainly fad and folly, and little else. And one need not be an ultranationalist to see some real problems in multinational companies that have revenues larger than the national income of some of the countries they operate in, and who make their decisions in headquarters that are far away from the countries that depend on them and far beyond the reach of these countries' governments.

But that these developments of the last ten years may not be to everybody's liking; that they may not all have been desirable or sound; and that not all their results may be enduring, does not alter two facts. They happened; and they have fundamentally affected economic structure, domestically (in every developed country) and internationally. Whether one likes them or not—and I personally have great reservations—one must ask the question: what explains them? One must assume that such far-reaching changes, which fly in the face of so much we considered "knowledge" in the field, must have their causes in major shifts in the structure of the economy.

II

The developments of the last ten years are, I submit, the first responses to the emergence of two new major "mass markets": (1) a mass-market for capital and investment, and (2) a mass market for careers for educated people doing knowledge work. Like all first responses, they were in all probability the wrong responses, or at best inadequate responses. But they were responses to real, new challenges—and the challenges will not go away.

Every economy, whatever its structure or its level of development, has three dimensions. One dimension is that of goods and services—their production, distribution, and consumption. It is the "here and now" of the economy. The second dimension is that of allocating resources to the future—the formation and investment of capital. Finally, in every economy, there is work, there are jobs, and there are careers.

Each of these dimensions, at whatever level the economy operates, needs some way of allocating resources. But it is only when an economy has reached the stage where it produces, in each dimension, more than the merest subsistence, that choices begin to become important. Only then can there be a "market" or "planning." And a "mass market" can only come into being when there is enough supply in each area for large numbers of people to make meaningful choices.

The "mass-market" level in respect to the first dimension, the production and consumption of goods and services, was not reached any place in the world until about 200 years ago. This was the essence of what the textbooks in economic history call the "commercial revolution" of the early 18th century—concentrated, of course, until the Napoleonic Wars, almost exclusively in England and the Low Countries. But even in England the well-being of the population was still measured, far into the 19th century, indeed until the repeal of the Corn Laws in 1846, by the fluctuations in the price of the standard loaf of bread. Even in England, in other words, the great mass of the people, until well into the 19th century, had very little choice in respect of goods and services and were not yet capable of forming a mass market. The market was determined by the supply rather than vice versa. In Japan, by the way, the price of the basic rice unit remained the standard measurement of economic welfare until the time of World War I.

The emergence of the mass market in goods and services in the 18th century explains why there suddenly arose a discipline of economics. For until there is meaningful choice, there can be no economics. But early economics, whether classic, neoclassic, or Marxist, did not produce a theory of money, capital, and investment, let alone a theory for the dimension of work and jobs. There was no need for this because supply in these two areas had not yet reached the level where there was significant choice. There have been markets for credit for many centuries. They go back all the way to the great fairs of the late middle ages. Still, as late as the early 1930's, when I was a young investment banker in London, then still the financial capital of the world, it was axiomatic that no more than one out of every 3 or 400 people, even in a highly developed economy, had enough savings to invest in anything but those financial necessities, life insurance and the mortgage on one's home. In other words, even in highly developed economies, as late as 1930, the great mass of the people, in respect of capital and investment, were determined by the available supply rather than capable of allocating it, to any significant degree, among choices. And all earlier economic theory, from Adam Smith through Marx to Marshall at the turn of the century, assumed that jobs were, by necessity, scarcer than available labor supply, whether this was being expressed in the "iron law of wages" or in far more elegant but essentially identical equations of "marginal utility."

Thus, economics—right through the 19th century—could treat capital on the one hand and labor on the other as being determined by the "real" economy, the economy of goods and services, and as having no autonomy and no dynamics of their own.

Keynes as a symptom

In this perspective, the development of monetary economics, culminating in the work of Keynes in the early 1930's bespoke a major structural change: the emergence of the capital and investment dimension of the economy into a true market system in which substantial numbers of people have choices. For Keynes—and even more for such post-Keynesians as Milton Friedman today—the capital and investment dimension is the "real" economy, with goods and services as dependent on it as, in earlier economics, capital and investment were seen as dependent on the goods and services dimension.

When Keynes published his great works, in the early 1930's, capital and investment were still very much a "specialty mass market" of a small minority, though one very much bigger than it had been only a few decades earlier. When the New York Stock Exchange announced around 1945 that it aimed at "mass ownership of shares," it had in mind raising the proportion of share owners from some 1 per cent to 3 or 4 per cent of the American population. The figure now stands at 80 per cent or higher. We have not "nationalized" capital, but we have "socialized" it in this country. One seventh of the population—30 million Americans—own shares directly; altogether they own almost two thirds of the total share capital of American industry. But the instrument through which the largest number of Americans own the "means of production" today are financial intermediaries—mutual funds and pension funds above all. They are the real "capitalists." These institutions are, so to speak, the "professional buyers" for the great mass of financial consumers. They are meant when the Securities Exchange Commission and the Stock Exchange talk of "sophisticated investors." And they owned, at the end of 1969, between 35 and 40 per cent of the equity capital of American business. As a result, financial ownership in this country is now distributed roughly in the same degree of equality—or inequality—in which consumption of goods and services was distributed in the early 1920's in the first great "mass-consumption" boom. The wealthy, that is, the top 20 per cent of the population, probably still own or control more than 35 per cent of the equity in American business. The poor, the bottom 10 per cent of the population, do not own any equity capital at all. The middle group, comprising some 70 per cent of the population, owns directly or through its intermediaries about half of the financial assets of American business.

An example of this development and of its speed is the pension fund of the American college teachers, the Teachers Insurance and Annuity Association. It manages the retirement funds for 300,000 people—for the great majority of the college teachers in the private colleges and universities and increasingly for teachers in state and municipal institutions as well. Founded in the early 1920's, it took Teachers 50 years to build up a pension fund of about $2½ billion in conventional annuities, invested mostly in bonds, mortgages, and other traditional life insurance investments. But less than 20 years ago, in 1952, Teachers added a "College Retirement Equity Fund," that is, a common stock mutual fund, and began to offer it to its participating college teachers. Today this fund manages a portfolio of common stock of well over a billion dollars—at which level, incidentally, it is still among the medium-sized rather than the large portfolio managers. And almost all the participants in the conventional fixed-payment annuity pension funds have also become common stock investors through the College Retirement Equity Fund.

What is rarely understood is that this is not just a quantitative expansion of "ownership." This is a qualitative change in the character of the "owners." They are not "capitalists." They are "investors." Their main stake in the economy is not through their investments but through their job and the income therefrom. What they can invest in the economy is, so to speak, "extra." They are not dependent on it. And, therefore, they can afford to take risks with the money. In fact, the most rational economic behavior for this group is to be a "speculator," that is, to invest for capital gains rather than for income and security. The middle income group, earning between $8,000 and $20,000 a year, is perhaps most conscious of the burden of our steeply progressive income tax and least capable of easing it through tax loopholes and tax dodges, precisely because its income is in the form of wages and salaries. Additional income from investments, therefore, holds out little incentive to them. "Growth," that is, the opportunity for capital appreciation, is worth almost twice as much for this group, in terms of real wealth, as additional income would be. And this group—which is, in effect, today's majority stockholder—is therefore singularly receptive to the promise of "asset management" and to "maximization" of the value of the stockholders' investment—and perfectly willing to accept considerable risk in order to attain capital gain and "asset maximization."

At first sight, this would not seem to apply to one large segment of these new investors, the pension funds. Traditionally, pension funds have been the most conservative investors, if only because they need a predictable income in the future to pay off their pensioners. But today's pension fund, especially the pension fund in industry, has been set up in such a manner that the one and only thing that is predictable is that the income needed in the future cannot be obtained from the contributions paid in the present, but must be supplemented by appreciable capital gains. In the first place, the typical American pension plan bases a man's pension on his income during the last five years of working life—and in a period of steadily rising wages, this means that contributions to a pension fund based on present wages will not be adequate for future pension needs. In addition, pensions increasingly are being adjusted retroactively for changes in the cost of living or in the wage level of the men still at work. In a period of rising wages and inflationary prices, future pension liability must, therefore, greatly exceed anything that present contributions can provide. Pension funds, therefore, in order to live up to the responsibilities under employment and labor contracts, simply have to invest for capital gains. The greater the inflationary pressure in the economy, the more will "speculative" behavior appear as the only truly "conservative" line of action for a pension fund; what used to be considered "conservative" behavior will appear as reckless and indeed irresponsible "speculation." The only way in which the typical pension fund of today can possibly hope to discharge its obligations is by investing for growth and capital gains, the only way it can possibly be true to its trust is by backing "asset management."

The new mass market for capital and investment thus must mean something very different by "performance" or by "value" than either the traditional "capitalist" or the "professional manager" mean by these words. These new expectations may appear "speculative" in traditional terms. But they reflect the economic realities of the new financial "mass consumer," the employed middle class. They may be unrealistic; but they are perfectly rational.

A mass market for careers

Perhaps more important in the long run—and far less seen by the general public as yet—is the emergence of the "mass market" in careers for educated knowledge workers, that is, for people with a college education. Only 40 years ago, at the eve of the Great Depression, there were very few careers in which one could really get paid for putting knowledge to work. Essentially these were the old professions, the ministry, teaching, medicine, and the law, to which, in the early years of the century, engineering had been added. The Yale graduates who were hired by Wall Street houses in the 1920's to be bond salesmen were nor hired for their knowledge; they were hired for their connections.

The supply of men and women with advanced education and prepared for knowledge work has increased almost 20-fold since 1920. But the supply of job and career opportunities for them increased even faster. In the last ten years the career choices for knowledge people have seemed to be practically limitless. In fact, that the young people today have to make choices of this kind—and without any real information—is surely a contributing factor to the unrest on the college campus. The young people are literally overwhelmed by all the opportunities that clamor, "take me."

The emergence of genuine choice for careers and jobs is so novel that there is no economic theory of a genuine "job market" so far. But there is the first sign that the old assumptions are being discarded by the economists. For the last ten years the "Phillips curve" has found increasing application in economic analysis. Named after an English economist who developed it, this curve relates inflation and employment. It attempts to identify the minimum unemployment needed in a given country to avoid inflationary pressures. It assumes, in other words, unlike any earlier economic theorem, that a shortage of men rather than a shortage of jobs is the "norm" of an economy, and that the labor market, far from being the corollary of the market of goods and services and of the market for money and capital, is in itself a major force molding the economy and shaping its other two major dimensions. But this is only a beginning. Of a true economic theory of the market for work, jobs, and careers, let alone anything comparable to Keynes' overarching model, there is so far not even the first sign.

Seventy years ago it was still true even in this country—as it is still true for the masses in the underdeveloped world—that most people had no choice of occupation. Even where jobs were plentiful rather than scarce, the job available to a young man—and his job from there on throughout his life—was largely determined by his father's occupation and his family's economic position. For the great mass of people this simply meant that son followed father—and in most cases, of course, that son followed father behind the plow. A few gifted or lucky or very enterprising ones could break out of this pattern—and in that respect, of course, American society has always had more mobility than any other. But even in American society, such mobility was the exception.

Today, no matter how unequally educational opportunities are distributed, the majority of the young have access to a college education, and with it to mobility and meaningful career choice.

Again, this is even more a qualitative than a quantitative shift. It is a shift from "looking for a job" to "expecting a career," from "making a living" to "wanting to make a contribution." The first employee indoctrination brochures, written by the early personnel men in the boom years of the mid-1920's when labor was scarce, usually started out with the question; "What does the job demand of you?" Today's recruitment brochures usually start out with the question, "What can you expect from the XYZ Company?" Where it was assumed, even at the peak of an earlier boom, that men were hunting for jobs, it is now assumed that jobs are hunting for men and that, therefore, they have increasingly to satisfy the values, demands, expectations, and aspirations of the knowledge worker. And, of course, the young knowledge worker is also the typical "investor" of the new mass-capital market. This explains in large measure what to so many observers, especially the older businessmen, appears as a total contradiction, indeed as a glaring hypocrisy, in his behavior. He loudly proclaims his "idealism" and his demand for a "career that makes a contribution"—yet in the next breath he asks the company recruiter: "and what stock options do I get and how much are they worth?" But given the realities of his situation, there is no contradiction in these two expectations to the young knowledge worker—or at least not more than most people can comfortably live with.

We have been told repeatedly during the last few years that the academician no longer owes "loyalty" to his college but instead to his "profession" or "discipline." But this is true of all other knowledge workers today. They have shifted from focus on the "job" to focus on the "career." The young academician is only the most visible of the lot, or perhaps only the one most visible to other academicians who write books. But, to see the full impact of this change, we had best go outside the United States, where mobility has always been taken for granted (way beyond its actual incidence indeed) to Japan.

In Japan, especially for professional and managerial people, a job has long meant "lifetime employment." The young man graduating from college was, so to speak, "adopted" into a "clan," whether a government ministry, a business enterprise, or a university, where he then stayed automatically the rest of his life. Both leaving one's employer and being let go by him were as carefully circumscribed as is divorce in the Roman Catholic canon law. One could leave for only one reason: to take over the business of one's old or ailing father. One could be let go only for very serious misbehavior or actual crime. These are still the official rules. But they are no longer being observed by the young, educated Japanese. They do indeed expect the organization they join to owe them "lifetime employment." But they themselves demand increasingly the right to move—with proper punctilio, of course. Increasingly they expect, what is even more of an innovation, a "career ladder" appropriate to their education and qualifications, rather than advancement by seniority. SONY was the first Japanese company to realize this. Skillfully blending Japanese traditions and the new values of the young knowledge worker, it combines the security of "lifetime employment" for those who want it with an open hiring policy for those who want to leave their old employers. It combines also pay and title by seniority with job assignments and career ladders according to qualifications and personal choice. This is probably the real secret of SONY's rapid growth and of its success against determined resistance to the brash newcomer by the whole force of the Japanese "establishment." It may be an example from which we, in the West, can learn a good deal.

The first responses

The structural changes in the economy, at home and abroad, can be understood as first responses to the wants and values of these new mass markets.

The "diversification mergers" were defensive responses aimed at shoring up an eroding competitive position in these two new mass markets. In essence they hardly differ from the similarly defensive mergers of the 1920's—except that what has forced formerly independent companies to band together, at heavy cost in management position and status, was growing deterioration and weakness in the capital and careers markets rather than in the market for goods and services. Many of the public announcements accompanying these mergers said so very clearly. The merger was undertaken because the combined companies expected to have "greater ability to raise capital," or "greater borrowing power," or "higher value for their shareowners"—as well as "greater career opportunities for our managers," or "more scope for our young people." Indeed, that a company can finance itself out of "retained earnings" is, in terms of the rationality of the new markets, a major weakness rather than a strength. It does not mean that a company is "secure" but that it is "stagnant": it isn't using as much capital as it should to achieve the fastest possible growth. It may offer income to its owners but not capital growth. It may offer jobs for middle-aged functionaries but not opportunities and careers to young knowledge workers (let alone tempting stock options). Many of the "diversification mergers" were, of course, only face-lifting; the new merged company is just twice as old, twice as bureaucratic, and twice as "undynamic" as each of the original businesses were before the merger. But in a good many other cases, the "diversification merger" did indeed provide the "synergism" the merger announcement talked about—but in respect of their performance capacity and ability to compete on the new mass markets of finance and executive talent rather in the older markets for goods and services.

By contrast, the "conglomerate" created by the "takeover merger" is an original, offensive response to the realities and demands of the new "mass markets." "Asset management" is simply another term for the design of a "product" that fits the values and demands of the new investor and of their new investing institutions. For "asset management" aims at increasing the value of a business by sloughing off the obsolete and unproductive old parts, and adding on to it new parts capable of rapid growth. In effect, the men who built the conglomerates by "takeover" tactics and financial manipulation presented themselves to the new investor as experts in obtaining the largest capital gain in the shortest period and in generating continuing capital gains and asset growth. These men understood that the new investors believe they can afford to take risks and want to. The conglomerate builders understood that "profitability," which used to mean "return on capital," has come to mean the "price-earnings ratio" for new investors, at least for the time being.

At the same time, the "takeover merger" also "packaged" for the careers market, though by no means to the same extent. It offers a small number of people opportunities at high risk for great gain—through stock options, for instance. It appealed to a good many young people who in their present positions within established, old solid companies had a "job" rather than a "career," let alone a chance to become an entrepreneur themselves.

Equally, indeed more than equally, the new entrepreneurs with their new "growth" businesses also represent a response to the demands and values of these two new mass markets. They appealed to two different "consumers" in the new capital market, with different preferences and values. They first sold to the "sophisticated investor"—that is, to the manager of the funds of the new financial mass consumer—the promise of rapid growth and with it of great appreciation of values. But many of them also realized that old, established, large businesses with limited growth opportunities can afford to pay very high prices for "glamor," that is, for adding smaller businesses with high-growth potential—it gives them marketability in the new mass markets. In many cases therefore the new entrepreneurs based their whole strategy on selling out to a "giant" as soon as their new venture had become profitable.

Above all, the new entrepreneurs represent a response to the new mass market for careers. They did not, as a rule, go after new graduates—they let somebody else train them and weed them out. They staffed their key positions—franchise owners, executives, editors, research managers, investment specialists, and so on—largely with men who had not found in their first jobs, whether in industry, in government, or in the university, the career opportunities that appealed to them. Indeed these new businesses would not have been possible had there not been a large and growing number of men whose demands for career opportunities the existing secure, safe, and even well-paid jobs with the old employers could not satisfy.

The multinational corporation and the off-shore investment trust are proof that these two new mass markets are not confined to America but, in varying degrees, are common to all developed countries. The US-based multinational corporation, which uses European capital to buy up European businesses, "packages investment" for the new foreign investor just as the "diversification merger" or the "takeover merger" packaged investment for the new American investor. It gives the European investor a chance for capital gain. It gives him marketability of his investment, whereas capital markets for most European securities in European countries are restricted to the point of being illiquid; and perhaps most importantly, it gives him public disclosure of performance and results, first imposed on United States corporations by the securities legislation of the 1930's—and thus the information for financial and economic analysis on which "sophisticated investors" and especially fiduciaries for other people's money must and do insist. It is no accident that the European and even the Japanese companies that have been going "multinational" have immediately had to adopt standards of disclosure which, while still inadequate by American criteria, constitute "indecent exposure" in a European, let alone a Japanese, business setting.

But, above all, the multinationals exploited the new "career markets" world-wide. Where, in Europe especially, traditional business still looks upon jobs as favors to be handed out, i.e., still acts on the assumption that even general managers are "hired hands" of whom there is an unlimited supply, the multinationals had to start out with the assumption that they had to create careers attractive to highly educated young men who had plenty of career choices. On this their success rests in large measure. In many cases the multinationals have become within a decade the industry leader in many countries—not by acquisition but by developing new businesses from scratch. This, it is now generally accepted, is a managerial rather than a technological or financial achievement—but this requires, specifically, ability to attract good young people, to hold them, and to allow them to go to work productively.

III

Takeovers, conglomerate-building, even new "growth" ventures have lost their bloom and their attraction. They became the first victims of the stock market collapse of 1969-70. In fact, it was the investor's disenchantment with these darlings of the boom that set off the stock market slide in the first place.

Such disenchantment was bound to happen—and this is not hindsight. It was quite clear from the very beginning that the investors who supported takeovers and went in for the new and untried "growth" ventures were naive to the point of being gullible (as well as greedy) rather than "sophisticated." It is plain silly to believe, as they did, that anything, and particularly a business, can grow forever at a high growth rate. If there is one thing certain, it is that any growth curve will flatten out, and that it will flatten out the faster, the steeper its initial rise. If there is one thing predictable in the case of a young and rapidly growing venture, it is a severe crisis—caused by the very growth of the business and the resulting strain on management's knowledge and ability to control. The faster a new business grows, the more severe will its "adolescent identity crisis" be. Until this crisis has been weathered—and that often requires changes at the highest management level—no one really knows if the business is viable, let alone if it can resume its growth.

Some of the conglomerates—though by no means all of them—have actually done quite well; some of the "asset managers" really managed assets and did succeed in making ailing businesses sound again and tired old businesses again able to venture, to innovate, and to perform. A large number—perhaps the majority—of the new "growth" ventures met and successfully overcame their "growth crisis." But a faith in magic is not maintained by statistical averages. And the faith in the conglomerate, in the creation of productive wealth by financial manipulation and "takeover," and in "growth" ventures that must succeed simply because they are in the currently fashionable industry—this was magic rather than "sophistication." It could not survive even the slightest shadow of doubt, let alone real setbacks.

As a result of this loss of faith, many of the conglomerates—all those that depended primarily on the "financial leverage" of a continuous rise in the price of their stock—are in deep trouble. The most spectacular performer of a few years back, Ling-Temco-Vought, has actually had to sell off some of its earlier acquisitions, and others will surely have to do the same. "Deconglomeration" may be the "in" word in the next few years. The most spectacular of the multinational "asset managers," Bernard Cornfeld's Investors Overseas Service, despite its almost 2½ billions in assets, ran into heavy weather in early 1970, as investors in its numerous affiliates and subsidiaries began to lose faith. Cornfeld himself had to resign.

Even the multinational manufacturing companies, though mostly strong, solidly-built companies with a firm base in their home market, face difficult times. Resistance against them on the part of national governments is growing rapidly; and this includes the United States government—Antitrust, the Congress, the Internal Revenue—perhaps the most parochial of major governments today.

The "inventory crisis" in careers

The mass market for careers will, at the same time, also go through a period of readjustment, will indeed face its own first "cyclical readjustment"—not because of economic conditions, not because of any cutback in governmental funds, but because of demography. The *supply* of college graduates is, obviously, heavily dependent on the number of babies born twenty-two years earlier. The *demand* in one important market-segment—teaching—is, however, heavily dependent on the number of babies born six to seventeen years earlier. Or to put it differently: young people have to be students and have to have teachers *before* they can themselves become available as teachers. If, therefore, there is a sudden sharp upturn in the number of babies born, there will be, a decade later, a sharp increase in demand over supply, with a resultant "inflationary pressure" on the career-market. And if there is a sharp down-turn in babies born, there will be, a decade later, "deflationary pressures." In fact, demographics (to which economists pay normally no attention) are the true dynamics of the "career market."

The demographic pressures on the American career market are just about to reverse themselves. Ten years ago the available new college graduates were still the "babies" of the very low birth years of the late 1930's. But since the post-World War II "baby boom" had exploded in 1946 and kept growing thereafter for ten years, the demand for teachers was, in 1960, at record height with a resulting excess of demand over supply—especially for women graduates—that created severe "inflationary pressures." Now, however, and for the next ten years, the situation will be the reverse: the supply will come from the very high birth years 1946-1957. It will be the babies of 1957, the peak year, who will be furnishing the graduating college class of 1980. The demand will come from the years after 1957, when the number of births at first remained even for a few years, and then, after 1961, began to fall quite sharply. Elementary school enrollment is already going down significantly. Junior high school and high school enrollment will start to go down in 1972 or

1973; college enrollment may well go down from its 1967 or 1968 levels; even to keep enrollment at its present levels would require a significant increase in the percentage of youngsters going to college. The careers market of the last ten years was heavily influenced by one of the sharpest jumps in the number of births ever recorded—almost 60 per cent, from the 2.3 million babies born in 1937 to the 4.3 million born in 1957. The careers market of the 1970's will be heavily influenced by a much less sharp but still almost unprecedented drop in the number of babies born between 1957 and 1967—a drop by one-fifth from 4.3 million to 3.5 million.

In some ways this is good news. For it means that other segments of the economy will not be as grossly undersupplied with educated people as they have been these last ten years—health care is the most important example. But to begin with, the total pressure on the career market will go down. Where there were several jobs hunting for a graduate these last ten years—when the number of graduates was still held back by the low birth rates of the 1930's and early 1940's—graduates and jobs will more nearly be in balance the next ten years, even if there is a boom in the economy. Graduates will have to learn again to hunt for jobs. Above all, however, there is what in the goods-and-services economy is called an "inventory imbalance." Teaching, having had the greatest and most crying shortage, has been heavily staffed with young people. This is true not only in elementary and high school but in college teaching as well (even though colleges, unlike elementary and high schools, can "stretch" the teaching supply by having large classes supplemented by "teaching assistants"—whereas public schools have class-size limits imposed on them by law). William Baumol of Princeton, the leading economist of American higher education, points out that in the 1960's American colleges hired five new and young teachers for every old teacher who retired or died. As a result, there will be a sharp drop in the proportion of teachers who, in the next ten and twenty years, reach retirement age or die—while the total number of college students will no longer go up, at least not sharply. Baumol foresees that the colleges will have to hire only three new teachers for every one who retires or dies in the 1970's—and only one for each vacancy in the 1980's! And in the lower schools the imbalance will be even greater—Geoffrey H. Moore, the Commissioner of Labor Statistics, and Conrad Taeuber, Associate Director of the Census

Bureau (as reported in *The Public Interest,* Spring 1970) calculate that there will be 4.2 million women graduates between now and 1980 who will, on the basis of past experience, look for a teaching job—and only 2.4 million openings for teachers in elementary and high schools.

But the present college students, undergraduate and graduate, are heavily, indeed preponderantly, oriented towards teaching—which is, of course, what the liberal arts degree concretely prepares for. They are not training for the jobs that will be there to be filled: health-care technology; professional and managerial jobs in local government (where most of the educated personnel entered during the Depression or early 1940's, and are nearing retirement age) and jobs in business, especially the highly technical "systems" jobs needed for work on environment and ecology. The students today complain that college tries to fashion them for jobs in the "military-industrial complex." They would have a much stronger case if they argued that they are being grossly misled—especially by younger faculty who, understandably, think that the last ten years were "normal" (whereas they were unbelievably "abnormal")—into preparing themselves for jobs that will not be available.

The shocked surprise in the spring of 1970, when the graduating class suddenly found out that they had to go out and look for jobs, may thus have been the first sign of a typical "inventory crisis"—which always takes everybody by surprise. Whatever the economic climate, the next few years will be years of sharp readjustment in the "careers market." The "career" boom of the 1960's is as much a thing of the past as the stock market boom in "takeovers," "conglomerates," and "growth ventures."

The new "asset managers"

But the new mass markets will remain the realities of the economy—in the United States as well as in all other developed countries. More difficult times, times in which sleight-of-hand and manipulation are not readily mistaken for performance, will only make greater demands on the development and management of institutions to serve these new markets, their customers and their producers.

The "fiduciaries"—investment trusts, bank-managed funds, and pension funds—will continue to grow as the dominant forces in the markets for capital and investment. Their importance is going to increase rather than decrease if the economy no longer can count on automatic inflation. For then the difference between the "sound" and the "unsound" investment, between the "growth" company and the stagnant or decaying one, between the well-managed business and the one run by chief clerks, really counts. And then the investor who, employed himself and working for a living, has neither time nor knowledge to pick investments for himself, needs an "informed buyer" far more than in boom times. He then has to insist on getting as his investment manager the "sophisticated investor" who so far, in a great many cases, has only been promised to him.

It is surely symptomatic that, when Investors Overseas Services ran into trouble in May 1970, the Rothschilds were offering to take over the floundering investment giant. For the Rothschilds have a unique record for turning into a permanent major institution a financial innovation when, as the Rothschilds had predicted from the first, the founder overreaches himself. This knack, demonstrated again and again in the 170 years of their history, is probably one explanation of the longevity of the Rothschilds' hold as a major financial power—outlasting by a good many years such earlier money dynasties as the Medici and Fuggers, not to mention the one-generation May-flies of finance (the Morgans, for instance) which America has so far spawned.

"Asset management" may altogether be more needed than before and become far more productive in periods of stable, let alone of declining, business. Then it is even more important, for the health of the economy as well as for the welfare of employees, communities, and shareholders that a company's assets be employed most productively. Then it is even more important that managements which, because of sloth, lack of imagination, or lack of competence, undermanage or mismanage the assets in their keeping be replaced by managements that can restore a company's capacity to perform.

This new "asset management" would, however, not build "conglomerates." Rather, one can predict, it will redeploy resources (abandon unproductive activities, for instance); it will bring in competent management; and it will then turn back a sound company to public ownership. Indeed, some such new "asset managers" have already begun to appear—some buying an undermanaged company with the intention of reselling it as a sound business, some working for the owners on a fee basis.

Altogether, the trends that characterized the financial markets of the 1960's are likely to become more rather than less dominant—even though actual forms may well change. The market is likely to be increasingly a market of "investors" rather than of "capitalists," and that means of people primarily interested in "asset growth" rather than in income. It will also, predictably, become an increasingly competitive market as buyers become better informed. It is already in all probability more competitive—and certainly far less concentrated—than the market for many goods and services. Companies needing money, and that means eventually every company, will therefore have to work on making themselves "marketable" to this new "mass market." It is most unlikely that concern for financial performance and financial marketability will decrease in the next ten or fifteen years. No matter what the New Left says (indeed no matter what economic doctrine the country accepts), concern for profit will predictably not go down during the next ten years, nor will concern with capital gains. The most we can expect is that a stable dollar and control of inflation may somewhat lessen the need of the fiduciaries, and especially of the pension funds, for continuing growth in the paper value of their assets.

Making the careers market work

In respect of the second of the new mass markets, the one for careers, a period of "inventory adjustment" will create a need for institutions adequate to serve the market. The market needs institutions which will play a role analogous to that of the "fiduciaries" in the mass market for capital and investment, i.e., intermediaries who are the "informed buyers" for the "mass consumer," the educated people looking for career opportunities, but also the "informed buyer" for the "mass producers," the employing organizations.

These will have to be "market institutions." The career and work dimension, especially as far as educated people looking for careers in knowledge work are concerned, is incapable of being "planned." Indeed the Soviet attempt to "plan" for individual careers and to direct young people accordingly has been the most dismal failure of Russian planning—far greater than the failure in agriculture. In the first place, the lead time needed to "develop the resource," that is, to train and educate for specific careers, is far longer than the time span for which we can project future needs; this time span, in general, runs to no more than six to ten years. In addition, "planning" here runs up against a simple law of mathematics: the impossibility of predicting the unique event (e.g., the toss of any single coin) from a large-number probability (e.g., the distribution of 1,000 tosses of a coin). And individual careers are "unique events" far more even than tosses of a coin. Finally, of course, "planning" careers means coercion and regimentation of the individual—and there is, fortunately, little reason to believe that our young educated people would enjoy this for their entire lives any more than they have traditionally enjoyed such "planning" for the—after all, limited—period of military service.

Yet we need to be able to provide information, guidance, and placement. This is particularly important as eight to twelve years hence we may well face another "demographic shift"—higher birth years are almost inevitable as the babies of the "baby boom" who now throng our college campuses marry and have children themselves. Beginning in the late 1970's the number of children in school will thus start to rise again. But, judging by past experience, this will be precisely the moment when the present guides of young people, especially faculties, will have adjusted their thinking to the changes in the opposite direction that are taking place now.

There are the beginnings of such market institutions. The last fifteen or twenty years have seen the emergence of the "head hunter," or "executive recruiter" (to give him his official name), the firm which finds executives in mid-career for new jobs. These firms are most active in business, of course. But there are some who find academic administrators, some specializing in clergymen to fill vacant pulpits, and others working exclusively as "head hunters" for hospital administrators. There are also, of course, the college recruiters of the major companies—something which was literally unheard of thirty years ago but which today is considered as much of a necessity in a large business as a research lab or a sales department. And there are a few firms which actually give career advice to the individual, and especially to the mature individual, who wonders whether he is in the right career path or whether he should change. Yet these are only first beginnings, comparable to the first investment trusts as they developed 40 years ago. They indicate what we need rather than themselves satisfying the need. Whether tomorrow's institutions in this new market will be "profit" or "not for profit," no one can predict. It may also not greatly matter. But the new "fiduciaries" who are the "informed buyers" in the careers market are likely to be a major "growth industry" of the 1970's.

Altogether the market for careers, in the long run, will become increasingly competitive. It will be increasingly important for the "consumer," the educated knowledge worker, to be able to make informed choices. It will be increasingly important for any institution to be able to attract the kind of people it wants and needs and to offer them what *they* need and look to. It will increasingly be important for any institution, business or nonbusiness, to be able to "market" its job opportunities. It will, therefore, increasingly be necessary for organizations to match the available job and career opportunities to the needs, demands, and wants of the customers. Perhaps one can sum this up by saying that, during the last fifteen years, there has been great emphasis on "manager development" aimed largely at making managers better capable to serve their organizations, whether business, government, or hospital. In the future, we are likely to balance this with growing emphasis on "organization development" aimed at making organizations capable of satisfying the aspirations of their "career customers," the knowledge workers.

And the new problems

These new markets will predictably generate new problems of public policy. Three in particular are already clearly visible—and for none do we have an answer. Indeed, with respect to none have we even tackled our homework.

The first and most *novel* of these problems will be the role and responsibility of the new financial powers, the "fiduciaries" of the financial mass consumer who are the real "owners" of today.

Majority ownership of America's (and Europe's) businesses will increasingly be controlled by people who are neither owners nor managers, but trustees. What should their role be? It can be argued (in fact it has been argued by some of the most thoughtful fund managers) that being trustees they cannot and must not interfere in the management of the companies they invest in. If they do not like a management, they can sell their shares in the company; but they have no authority from anyone to exercise control. But if these institutions do not exercise control, who does—or can? Management either is uncontrolled and uncontrollable, or the policing function falls to the "takeover" entrepreneur. Clearly neither solution is acceptable. Yet no new one is in sight.

But the problem can also no longer be avoided. Indeed it has already been raised—though in the least expected form. "Nader's Raiders"—the young lawyers who work under and with Ralph Nader of automotive-safety fame—raised it when they asked, in the spring of 1970, that foundations, endowments, and other trustees withhold their proxies from General Motors management and vote their shares instead for a series of changes in the company's board, changes in policies and management which "Nader's Raiders" hold to be in the public interest. Almost at the same time, though of course in a totally unrelated development, the Antitrust Division of the Department of Justice filed suit against a very large fiduciary manager, the Continental Illinois Bank in Chicago. What the Antitrust Division charged was that the Bank's practice—and every major bank engages in it—of having different officers of the bank sit on the boards of directors of competing companies was in violation of antitrust even though these directors did not represent "ownership" but simply the shares held by the bank in trust for very large numbers of individual beneficiaries. That, in effect, Antitrust attacks as illegal precisely the very exercise of control—and the same interference—the absence of which "Nader's Raiders" castigate, does not affect the importance and seriousness of the issue, nor does the fact that nothing in our economic experience is much help deciding it. But it will have to be tackled—and soon. It will, indeed should, become a major issue of public policy.

Multinational company vs. national state

The second and most *important* problem—important in terms of its impact—will be that of the multinational business.

Professor Perlmutter's prediction that, in another fifteen or twenty years, the world's manufacturing production will be in the hands of 300 mammoth multinational companies, while widely quoted, represents an extreme rather than the most probable trend. But it is not rash to predict that, within every developed noncommunist country, a fifth or a quarter of total manufacturing output will be produced by companies that are "multinational" in their operations. Indeed, this is reality rather than forecast in ten of these countries (the United States, Great Britain, Canada, Germany, Italy, Holland, Belgium, Switzerland, Norway, and Sweden), is fast becoming reality in a ninth (Japan), and is some distance from accomplished fact in only two (France and Brazil).[1]

This means that in the developed countries a very big part of the economy is subject to decisions made beyond the reach of the national government. But it also means, conversely, that governmental decisions in most developed countries—France and Canada being the only important exceptions—have impact far beyond the country's own borders through the effect they have on the multinationals headquartered in that country. American antitrust regulations, tax laws, and restrictions on trade with Communist powers are held by government authorities to bind the subsidiaries and affiliates of United States-based companies everywhere. No other country is quite so openly nationalistic. But other governments, too—especially those of the big countries such as France, Germany, and Japan—like to look upon "their" multinationals as instruments of their own economic policies in the world, while at the same time bitterly resenting that the subsidiaries of "foreign" multinationals on their soil are, in some measure, beyond their complete control.

This is not a problem of "capitalism." Indeed, the same ambivalence characterizes economic relations within the Soviet bloc. Quite clearly the multinationals which the Russians have been trying to build throughout Eastern Europe are primarily resisted for political reasons. They

[1] Of these, incidentally, the United States, Holland, Switzerland, and Sweden are "headquarters countries." In Germany, ownership of industry by companies based abroad—mainly United States, Dutch, and Swiss—and ownership of businesses abroad by German companies roughly balance. In Japan, where the "joint-venture" predominated—i.e. a partly foreign-owned company doing business in Japan in partnership with a Japanese company—ownership or co-ownership of businesses abroad by Japanese companies is growing so fast that Japan may soon also be in "ownership balance." Great Britain, Canada, Italy, Belgium, France, and Brazil are far more "owned" than "owning."

remove a part of the Polish, Czech, or East German economy from the decision and control of the Polish, Czech, or East German governments. Multinationals, whether "capitalist" or "Communist," put economic rationality ahead of political sovereignty.

De Gaulle's opposition to the "multinationals" was therefore not "anti-American." He also opposed non-American multinational attempts—e.g., that of the Italian Fiat company to merge with Citroen, France's ailing automobile manufacturer—as vigorously as he opposed the Americans coming in. He forbade French companies to become multinational themselves and to move beyond France. Indeed, de Gaulle's insistence on the congruence of political and economic sovereignty was completely consistent and was the only rational policy for the problem worked out so far any place by anyone.

It was also a total, resounding failure. A larger part of the advanced sectors of French industry—computers or pharmaceuticals, for instance—is controlled by foreign multinationals than is the case in any of the large developed countries other than Canada. French capital is more lavishly invested in multinationals than the capital of any other of the "majors." *Not* in French-based multinationals making their decisions in Paris—there are none, thanks to de Gaulle—but in the shares of foreign-based, i.e., American, Swiss, Dutch, and Swedish multinationals. And at the same time, there is no country where so many of the ablest young executives, researchers, and managers work for foreign-based companies. What defeated de Gaulle, in other words, were the pressures and preferences of the two new mass markets, especially perhaps that of the new "consumers" of the career market, the young, educated knowledge people.

Yet de Gaulle, with his usual clarity, at least saw the problem. The multinational corporation is by far our most effective economic instrument today, and probably the one organ of economic development that actually develops. It is the one non-nationalist institution in a world shaken by nationalist delirium. It is not a political institution itself and must not be allowed to become one. Yet it puts economic decisions beyond the effective reach of the political process and its decision-makers, the national governments. This may well be exactly what we need to de-fang the nationalist monster. But national governments and their organs, whether legislative or executive, are unlikely to see it this way. And how we accommodate this tension between economic rationality and political sovereignty in the next few years will have a tremendous impact on both the economy and the working of government.

When concentration is competition

The third and most *difficult* problem posed by the emergence of the new markets and the new entrepreneurs are those of concentration and competition. In the other two areas we have to find new answers. In respect to concentration and competition, we have to unlearn old ones. And that is far more difficult, especially as the old ones have been held with almost religious fervor and have almost become sacred chants for large groups of economists, politicians, lawyers, and businessmen.

Two concepts have guided our approaches to the problems of concentration and competition for many years: "concentration of manufacturing assets" and "concentration of market power." The measurements developed for these two aspects of economic concentration are widely accepted as giving us in conjunction both an X-ray photograph of the bony structure of our economy and reliable guides to diagnosis and treatment. But the first measurement is becoming unreliable, while the second one has become misleading.

For a long time "concentration of manufacturing assets" remained fairly constant. But according to the antitrusters, it took a tremendous jump upward in the last twenty years. In 1950, the 200 largest "manufacturing companies" controlled 40 per cent of the country's manufacturing assets. In 1970, the top 200 control 60 per cent—the biggest increase in economic concentration ever recorded in this or any other country.

The odd thing, however, is that this tremendous concentration has not been accompanied by any increase in concentration in economic power in any single market for goods, that is, in any single market in which manufacturing companies operate. In most of these markets, concentration has probably gone *down* during the last twenty years. In market after market, new companies have challenged the big old companies and have taken away from them a piece here or a piece there of their traditional business. This is true whether we speak of book publishing or of pharmaceuticals, of building materials or of retail sales.

"Manufacturing assets" no longer define the concentration in the American producing economy. What is counted in this rubric includes assets shown in the balance sheets of American-domiciled businesses wherever the assets may actually be, whether within or without the United States. In 1950 these assets were almost exclusively within the United States. Today, however, most major American companies are multinational, with at least 20 to 30 per cent of their production and assets outside of the United States. Therefore, at least one quarter of these 60 per cent —that is, 15 percentage points—should be subtracted from the official figure, which would bring the rate of concentration in American manufacturing down to 45 per cent.

At the same time, however, what counted as "manufacturing companies" in 1950 were largely companies that were actually manufacturing. To be sure, General Motors even then owned one of the largest finance companies, the General Motors Acceptance Corporation; but its assets were a very small fraction of total GM. Today, as a result of "diversification" and "takeover" mergers, a very substantial number of companies which are still counted as "manufacturing companies," actually have very large assets—in some cases the majority—outside of manufacturing, in service businesses and above all in finance. And in finance, "assets" are not really assets but are essentially "liabilities," that is, money borrowed to be lent out immediately. During these past twenty years, whenever a manufacturing company merged with a financial company it acquired on its balance sheet financial assets very much larger than its own "manufacturing assets" had been—even though in terms of profitability, let alone of economic power, the manufacturing company may well have been the bigger one. These financial assets are, however, from then on considered "manufacturing assets" in the figures. When Control Data acquired Commercial Credit on December 31, 1967, it had manufacturing assets of $470 million. Commercial Credit had assets of $3 billion—the total of both is, however, now considered "manufacturing assets," since Control Data was legally the acquiring company. If the biggest of all attempted "takeovers" had gone through—that is if Leasco, a computer service company, had succeeded in taking over Chemical Bank-New York Trust Company—Leasco assets of less than $800 million would have been augmented by Chemical Bank assets of $9 billion, with the combined total of almost $10 billion all counted as "manufacturing assets." We have therefore to deflate the official figure for "manufacturing assets" by at least another 10 percentage points to take out assets that should not have been counted as "manufacturing assets" at all. In other words, in terms of true "manufacturing assets" in the country, the 200 largest companies today almost certainly have a smaller share of manufacturing industry than they had 20 years ago.

This would bring the two sets of figures—manufacturing assets and concentration of market power—back into alignment. Yet, clearly, the conclusion that there has been no "concentration" is not plausible. For while there has been neither greater concentration of American manufacturing assets nor greater market-concentration, the diversification and takeover mergers—and the multinational expansion—*have* clearly produced a considerable concentration in decision-making power. They have led to very large businesses, acting in many areas and countries but nonetheless incorporated in one legal entity and directed by one top management.

The result, however, is often *increased* competition, even in the goods and services economy. And the result is almost always increased competition—indeed deconcentration—in the capital and investment economy and in the work and careers economy.

The Control Data-Commercial Credit merger increased concentration in neither the computer nor the installment-paper market. On the contrary, it made these markets more competitive by strengthening what had been the "underdogs" in both. An even more telling example is the acquisition a few years back of Folger, a rather small regional coffee blender, by Proctor & Gamble. This clearly added to Proctor & Gamble's bigness. Since P&G is also a leader in processed foods, the new acquisition also added, albeit not greatly, to its market share in the "processed foods" industry and thereby to industrial concentration. But with the resources of Proctor & Gamble behind it, Folger could reach out for national distribution in the coffee market. Because the national coffee market had, for years, been dominated by a few brands in a typical "oligopoly" pattern, Folger's acquisition by Proctor & Gamble therefore also meant significant *deconcentration* in one important market. What then were the "real" consequences—concentration or deconcentration?

We may well be drifting towards a situation in which leadership and concentration in one market—that of goods, of capital, or careers—is the "countervailing force" for competition and deconcentration in one or both of the other markets. Surely it is not without relevance that the most common criticism of "multinationals," whether the Americans in Europe or the Europeans and Japanese in the United States, has been that their size enables them to indulge in "excessive competition."

The pressures towards this kind of concentration, which is so very different from what the term has implied traditionally, will increase rather than decrease. Technology is pushing in that direction, especially in the materials and chemicals industries. Technology is forcing Du Pont, traditionally primarily a producer of chemicals for the textile industry (e.g., synthetic fibers), to go into pharmaceuticals on the one hand and composite materials including new combination metals on the other. Technology has already forced the two big can companies, twenty years ago producers of a single product, the tin can, to become manufacturers of "packaging" which includes plastics, glass, paper products, and so on. And this, in turn, forced the largest manufacturer of paper-based packaging, Container Corporation of America, as said above, to merge with a retail and mail order chain, Montgomery Ward, to obtain enough financial and management muscle to stand up to the new packaging giants. Another powerful force moving business towards concentration will be concern with the environment. Purity of heart by itself will not clean up the environment, whether we talk of air, water, the open spaces, or the city. It will require massive "systems" effort in every area, that is, companies that can mobilize major technological and economic resources across a wide variety of skills, disciplines, technologies, and markets.

But above all, the pressures of the new mass markets, the mass markets for capital and investment, and for jobs and careers, should push for continuous concentration along the new lines—that is, along the line of diversification in industry and diversification in geography.

We will, therefore, have to think through what kind of "diversification" is desirable, productive, and rational, and what is simply financial manipulation and empire-building.

What kind makes the economy more open, more flexible, more competitive, and what kind furthers concentration and monopoly? Which one creates enterprises that are more manageable and perform better, and which one creates managerial monstrosities?

What we should want is reasonably clear. We want diversification rather than diffusion. We want federalism rather than either centralized tyranny or dispersion. We want "asset management" rather than financial manipulation. But into which of these categories a given structure falls, is by no means clear. Indeed it is not even clear to the antitrusters, who are sharply split between those who accept and indeed welcome "conglomerates" as leading to increased competition, and those who bitterly oppose them as producing increased concentration. This issue predictably will be one of the main concerns of the next ten years, in the United States as well as abroad. That there is no "right" decision is not so important—there rarely is for problems of this kind. But that the old and accepted concepts and measurements are no longer appropriate, is going to make the going rough. And that we will have to learn to "trade off"—that is, to balance concentration in one economic dimension with competition in another—goes against the grain of decision-makers, whether economists or politicians, businessmen or bureaucrats; they all, understandably, resent and resist such complexity.

A new reality

The economic developments of the last ten years signify more than a change in economic structure. They changed economic reality. This will require new thinking and the sloughing off of a great many traditional concepts, ideas, and policies in respect to "monopoly," "concentration," and "competition," for instance, and in respect to the relationship between the world economy and the nation-state. It will require the development both of new theoretical understanding and new policy concepts. For so far we have no economic theory that embraces or even connects the three dimensions of the economy and thus integrates the new "mass markets" of capital and investment and of careers with the old mass market of goods and services, prices and productivity. The specific developments that characterized the 1960's may well have been temporary phenomena, never to recur. The developments of which they were the first expression and the visible symptom have only begun.

CAPITALISM AND THE ECONOMISTS

4

On the limited "relevance" of economics

ROBERT L. HEILBRONER

RELEVANCE is a word that makes professors of economics wince these days. There was a time when I could initiate a class into the mysteries of diminishing marginal utility, explaining why the man in the Sahara Desert would not be willing to pay as much for the third pint of water as for the second, confident that when the hands went up it would be because someone was convinced that he ought to pay more, because his *total* utility was greater after three pints than after two. Today when the hands go up, I know what's coming: "That's clear enough, but I don't see how it's relevant."

Is it relevant? Have the refined figments of economics anything to do with ghetto life or the behavior of corporations or the military-industrial complex? It is easy to sympathize with the student who chokes on marginal utility while the world seethes outside the classroom. Yet, I do not think that this first attack on the relevance of economics counts for much. For, as his instructor quickly explains, these rarified concepts are essential if economics is to penetrate to the fundamental elements of the social universe. They are as necessary to economics as the abstractions of time and space are for physics. Indeed, if it could not conceive of an abstraction like marginal utility, economics could not claim to be a social *science*.

Moreover, as the instructor delights in pointing out, some of these remote conceptions contain unexpectedly sharp cutting edges. The rationale for progressive taxation, for example, uses the argument that successive dollars of income, like successive pints of water in the Sahara, yield ever smaller increments of enjoyment to their recipients. Pure competition, another hopelessly "unrealistic" concept that is a favorite target of freshman scorn, turns up, of all places, as the starting point for Marx' analysis of the dynamics of the capitalist process. Indeed, by the time a zealous instructor is through, the danger is that the shoe will be on the other foot, and the class will have been persuaded that "irrelevance" is nothing but the ill-considered objections of those who have not yet mastered the subject.

Yet, it must be clear from the title of this essay that I am not among those who are so persuaded. For behind the question that our student puts, is a deeper issue to which the standard reassurances do not always address themselves. It is that economics exists for a purpose, and is meaningful and useful only insofar as it serves that purpose. The purpose is to enable us better to comprehend the structure and tendencies of the economic order—that is, the institutions and activities that affect the production and distribution of wealth. Insofar as economics gives us this understanding, no matter how abstract or far-fetched it may seem, it is undeniably relevant. But insofar as it fails to do so, no matter how elegant or scientific it may appear, it is not.

I do not think an instructor would take issue with this criterion of relevance. Rather, he would argue that economics meets it very well. Does not the enormous demand for economists, he would ask, imply that the discipline has something to offer for which hard headed businessmen are willing to pay? Does not the recent creation of a Nobel Prize in economics testify that it is an activity of high public importance? Is not the flattery of imitation paid by other social sciences to the techniques of economics similar testimony from the world of scholarship? What further proof of relevance would one want?

The answer I wish to give is not to denigrate what economics has done. It is to stress what it has failed to do. Obviously, economics throws a good deal of light on the workings of the economic process. But the light is curiously uneven. Some parts of the social machinery are brilliantly lit. Other parts are left shadowy or totally dark. And this uneven lighting, I would maintain, is not merely due to the fact that, like all sciences, economics is more highly advanced in some areas than in others. It is the consequence of the kind of illumination that conventional economic theory sheds. In other words, I believe there are aspects of the economic system that economics, as it now exists, *cannot* light up, no matter how brightly its lamps may burn. For someone who wants to inspect those parts of the machinery, economics today is as useless as a searchlight fixed in the wrong direction.

The irrelevance of economists

We shall turn shortly to the problems inherent in the nature of economic theory itself. But first we must take up a touchy subject that cannot be sidestepped. For one reason that part of the social machinery is shrouded in darkness cannot be attributed to the failings of the discipline itself. It lies rather in the fact that economists themselves are at fault for having failed to turn their torches on important issues and problems of our economic order.

Let me give a few instances of this disregard of certain economic problems. One area of nearly Stygian blackness is to be found in the growth of American economic power abroad. Between 1950 and 1969, American overseas investment increased from $12 billion to over $65 billion, giving rise to the familiar charge that American imperialism has become a dominant force in international economic (and perhaps political) affairs. Astonishingly, however, a student of international events who depended for his knowledge of current trends on the professional journals would have remained totally unaware of this portentous development. The statistics of foreign investment were, of course, reported. But I am unaware of an analysis of the politicoeconomic consequences of this economic penetration in the professional literature. Our knowledge of the extent, nature, or effects of American hegemony in foreign investment has been gained almost entirely from the work of Marxian economists or from that of journalists such as Servan-Schreiber.

Another striking evasion is to be found in the professional examination of that central abstraction, the firm.

There is a vast literature on the firm, but studying it would leave one curiously uninformed about many aspects of its real-life counterpart, the corporation. Reading professional journals one would not know that corporations affect national policy; that they are centers of powerful propagandistic efforts; that they practice fraud, chicane, and misrepresentation; that they dominate the social and political life of many smaller communities in which they are located, etc. Economists would no doubt offer a number of reasons why these aspects of "the firm" were left unexamined, principal among them being that these were not "economic" questions. But the fact remains that they *were* left unexplored. So far as conventional economics was concerned, these issues simply did not exist.

This roster of ignored subjects could be easily expanded. I do not know of any treatment in the professional literature of the size and distribution of the benefits of the war economy by socioeconomic groupings; of any interest in the question of the means by which private wealth and income are preserved from one generation to the next; of any analysis of the benefits accruing from "public" goods by income strata. Even poverty and ecological damage, now fairly common subjects for professional examination, were not issues originally unearthed by economists, although both problems would seem to lie directly beneath their gaze.

This is not to say that nothing has been written on these matters. Here and there an economist has looked in these directions, at aspects of the social machinery that were formerly obscured. Yet I do not think that many will dispute my contention that these are not the kinds of questions that crowd the pages of the economics journals, or that the attitude of most economists has been to avoid rather than to seek out perspectives of this kind.

I would feel somewhat more uneasy in my contention that economists were markedly one-sided in their outlook were not my belief bolstered by the support of an impeccable observer, George Stigler, one of the most respected economists of the conservative Chicago school. In a well-known essay, "The Politics of Political Economists," Stigler has written that "the professional study of economics makes one politically conservative." The reason for this is "the effect of the scientific training the economist receives. . . . It becomes impossible for the trained economist to believe that a small group of selfish capitalists dictates the main outlines of the allocation of resources. . . . He cannot unblushingly repeat slogans such as 'production for use rather than for profit.' He cannot believe that a change in the *form* of social organization will eliminate basic economic problems."[1]

Professor Stigler's point is clear: if economists tend to be conservative, this is because it is the more *intelligent* thing to be. Without conceding that point (for I am by no means convinced by Stigler's examples of the things that economists cannot believe), I would like to raise another, less elegant, explanation for the phenomenon. It is that economists are conservative because they tend to be located in the upper echelons of the pyramid of incomes. The demand for economists has now raised their salaries until (as of 1967) the median pay of an associate professor was $14,000, that of a full professor $18,000, and of a "superior" full professor $21,000. This scale of salaries, quite aside from the royalties, consultation fees, foundation grants, etc. which are the common complement of professional pay, suffices to place associate professors in the top 15 per cent of all taxpayers; full professors in the top 5 per cent; and "superior" full professors in the top 2 per cent.

In his article Stigler raises venality as a possible cause for the conservatism of the profession, and dismisses it with the comment that "current rates of pay for good economists are much below what I would assume to be the going rate for a soul." But venality is not the issue at stake. When we seek to explain the prevailing political outlook of other groups in society, it is first and foremost to their place in the socioeconomic spectrum that we look. Thus, if economists are conservative, one *prima facie* reason would seem to be that they simply share the conserving attitudes we find in the top echelons of all societies.

To this one should no doubt add that special contemplative approach of the scholar which in itself tends to militate against "radicalism." But the first consideration remains to be faced: economists fare very well in this social order; why should they search for problems that might ruffle that serene temper on which conservatism always rests? The answer, to come out with it flatfootedly, is that they do not, with the result that the picture of the economic order that conventional economics gives us is

---

[1] *Quarterly Journal of Economics,* Nov. 1959, reprinted in *Essays in the History of Economics* (Chicago, 1965) pp. 52, 59, 60.

not one which might emerge if every portion of the mechanism were illumined as brightly as every other portion, but a partial and selective view from which much that is unpleasant or disturbing has been, no doubt unconsciously, passed over.

Let me now pass to more elevating matters. For there is a second and more interesting reason for the unreliable illumination that economics affords. It is (to continue our metaphor) that the light of economics only seems to work when the machinery of the social system is turning over smoothly, whereas any failure in the social mechanism short-circuits the beam that economics throws.

This will take us into a quick tour of the electrical arrangements. As anyone knows who has ever looked into an economics journal, economics is now a very complicated subject, virtually inaccessible to someone without a fairly thorough grounding in mathematics and statistics. Yet what is really significant about the discipline communicates itself immediately to the most casual browser. It is that economics bases itself on the methods of "real" sciences. In the severity of its language, its omnipresent formulae, the recurrent use of the word "model," there is clear advertisement that the paradigm of elucidation that economics follows is patterned as closely as possible on that of the physical sciences. In particular, economic methods stress rigor of procedure and proof, and a studied absence of value judgments—an approach that finds its ultimate expression in the depiction of the economic process as a series of interconnected mathematical equations.

Now what is astounding about economics—or more accurately, about society—is that this mathematical paradigm actually finds its application to social reality. That is, economics can indeed construct algebraic (or geometrical) analogues of social situations that explicate social action with the irresistible force of a logical demonstration, an achievement before which other social sciences pale with envy.

Because this point is very important, let me explain it in more detail. Economics derives its extraordinary explicative power from two sources. First, unlike its sister social sciences, it carefully limits its concerns to the *quantifiable* aspects of the social process. It is not with the sweat on the laborer's brow or the pleasures on the consumer's palate that economics deals, but only with the units of output the laborer produces or the wages he earns, or with the amounts the consumer expends or the income he enjoys. Thus, economics claims for itself a portion of the social universe that is peculiarly amenable to mathematical representation—a claim that immensely facilitates its treatment of the universe in a "scientific" way.[2]

Second, economics achieves its scientific capability because it discovers two kinds of social behavior at the heart of the economic process, both of which lend themselves with surprising ease to replication through mathematical equations.

The first of these behavior patterns is given its clearest expression in the crossed "curves" familiar to every freshman as the Supply and Demand diagram in chapter one of his textbook. As the freshman quickly learns, the curves describe the fact that buyers and sellers, interacting on a marketplace, will spontaneously discover that there is one "equilibrium" price that clears the market—a price at which the quantities offered for sale will exactly equal those that are sought by the other side.

Entranced by the beautiful simplicity of this truly remarkable diagram, our freshman may not reflect on the still more remarkable property of the social mechanism on which it rests. It is that one and the same economic stimulus—for example, a rise in prices—will result in *opposing* behavior reactions. As prices rise, buyers will reduce the quantities they are willing and able to take from the market, while sellers will increase the quantities they seek to unload on it. From the interaction of these opposite behavioral patterns there then emerges the astonishing possibility of describing a social process by means of two different equations, and the even more astonishing possibility of predicting the outcome of this social process by solving these two algebraic expressions.

A second, although related, characteristic of society offers the economist a second basis for translating social action into mathematical formulae. It is that much economic activity, especially that involving production, can be described as "maximizing subject to constraint." By this the economist means that the profit-seeking firm or individual does not have the capability of increasing its revenues without limit simply by adding more and more inputs. Sooner or later, the hard facts of scarcity, technological barriers or organizational inertias place boundaries

---

[2] And what about marginal utility? Economics has experienced no end of trouble with utility precisely because enjoyment is not cardinally quantifiable. But it gets around the problem by *ranking* one enjoyment as greater than, equal to, or less than another.

in the way of unlimited profit-seeking, usually in the form of ever more steeply rising costs.

So much is no more than common sense. But now follows the achievement of economic science. By calculating the "shape" of the constraints imposed by technology, scarcity, or whatever, the economist finds that he can reproduce in a set of equations the outcome of the interaction between the force of profit-maximizing and the constraints of the physical and institutional world, just as a physicist can give mathematical representation to the forces and resistances of the physical universe. Hence, at least in theory, the economist can *predict* the behavior of the profit-maximizing firm in a model that once again displays the extraordinary property of describing social behavior in mathematical terms.[3]

Thus in the eyes of the theoretical economist, the unusual attribute of economics—its claim to relevance, if you will—lies in its ability to portray important aspects of the social universe in terms of the same kind of determinate models that we use to depict certain processes of the natural universe. What is more, there is no doubt but that this fantastic claim is partly true. There *are* relationships of an opposing-behavior or force-constraint kind that hold with sufficient regularity to enable us to use economics with considerable accuracy as a means of explaining or predicting many economic events. As I have said, the practical usefulness of economics is without question, and much of it derives from this "scientific" base.

Economics and the real world

There is, however, a problem. It lies in the fact that the premises of opposing-behavior and force-constraint on which so much of the scientific model rests are *behavioral assumptions,* and these assumptions, while valid enough at the time to give the scientific model its usefulness, are not valid all of the time. Behavior, in a word, contains an element of indeterminacy that plays hob both with real social processes and with the accuracy of their representa-

---

[3] Let me add that these two characteristics of opposing-behavior and force-constraint are not the only bases for scientific economics. Economics also relies on statistical relationships that express "typical" behavior patterns, such as the propensity of individuals to save or spend at different levels of income. Nevertheless, the two relationships discussed above are central to most economic theory.

tion in mathematical models. For when behavior departs from the pattern on which economic theory is built, the illumination—the relevance—of that theory suddenly disappears.

But how indeterminate is behavior? Clearly, it is not so capricious that Macy's cannot count with considerable assurance on a positive response from buyers when it lowers its prices. In "ordinary" circumstances the behavior of buyers and sellers and producers is sufficiently like that on which economic theory builds its edifice so that economics enables us to make predictions and analyses of considerable accuracy. But conditions, alas, are not always ordinary. Indeed, in precisely those times when we most want the assurance of theoretic reliability—when, for example, times are unsettled, or the economy seems poised at the edge of a boom or bust, or when expectations are particularly labile (such as is the case right now)—the dependability of the behavioral premises rapidly deteriorates.

As it does, the determinacy of the scientific model of society deteriorates as well. In the supply-demand situation, for example, the presumption, as we have mentioned, is that buyers will respond to higher prices by curtailing their activities, while sellers do just the opposite. But once an uncertainty of outlook is introduced into the picture, this assumption disappears. Now buyers may respond to a rise in price by rushing into the market before (they think) prices go higher; and sellers may respond by curtailing their offerings, in the hope of selling them more profitably tomorrow. In that case it is impossible to tell where the "equilibrium" price and quantity will settle, because we do not even know, from moment to moment, where the "curves" themselves are located or in which way they incline. That is why the marvellously sophisticated computer models of the economy, which perform well as long as things fundamentally conform to the behavioral patterns of normality, lose their predictive power totally when a sharp change in the winds of opinion brings about a change in economic habits. It explains also why the government which can build elaborate models of the economy for normal periods, detailing with considerable accuracy the effects of small changes in monetary or fiscal policy, cannot tell in advance how a major change in policy will affect the economy because it cannot know how such a change will be "interpreted" by the public.

Although it is less dramatic, a second weakness in the economic depiction of reality further undermines the scientific pretensions of the economic model. This is the assumption that we can actually describe the economic forces that bump up against constraints. When the beginning student first works with this problem, he finds most interesting the determination of exactly where along its frontiers of possibility a firm will choose to locate itself. But the more bemusing question is a much simpler one: will the firm in fact be operating on its "frontiers," or will it be content to get along with something less than the "maximum" profit that the model assumes as its goal. And the problem is not merely one of deciding on the degree of fineness of economic accuracy in prediction. For if firms do not maximize, we have no way of knowing what economic path they will follow, and the whole scientific reproduction of that part of the social universe collapses forthwith.

The question then comes down to an empirical one: do firms (or individuals) actually maximize their profits (or utilities) or whatever? What is embarrassing is that we cannot tell whether they do or not. In part, given the complexities of technology and the inertias of organization, this is the result of the fact that it is difficult to determine at any moment what "maximum" performance for a firm should be. More important, maximization becomes an indeterminate guide because it is impossible to translate into a prescription for behavior over time. Given the innumerable factors that may affect its sales, a firm may rationally seek to maximize its profits over "the long run" by increasing or decreasing its investment, by pushing product A or product B, by risking an antitrust suit by aggressive actions or by guarding against one by passive behavior.

The problem is thus that profit-maximizing is a rule of conduct that it is impossible to specify or detect in the real world. Hence it is virtually impossible either to predict future behavior or to evaluate existing behavior in the terms of its congruence with an objective standard.[4] The analogy is that of a physicist who was unable to state whether an object dropped from a tower was actually

[4] The exception is for firms competing under conditions of pure competition. Here there is an unambiguous rule of profit-maximizing that governs their immediate behavior. But such firms are themselves the exception in the world of imperfect competition that we find in actuality.

falling at the rate predicted by the law of gravity. In such circumstances, the law could no longer serve as a basis for predicting events in the universe. In similar fashion, our inability to state whether or not corporations (or individuals) are in fact operating on their maximizing frontiers vastly reduces the relevance of the scientific model by which we seek to elucidate the workings of the economic universe.

Economics and social change

What we have seen so far is that economics lacks relevance insofar as its practitioners are loath to apply it to "radical" critiques of the system, and to the extent that the real world defies the regularities of behavior on which the economic model of reality is based. But there is, as well, a third problem to be faced. It is that economics, by its very insistence on dealing only with the quantifiable elements of social activity, thereby distorts and misrepresents the activities it seeks to explain. Economics deals with the economic order as if it were only a mechanism for the generation and allocation of goods and services. But the fact is that the economic order is also and inextricably a mechanism for the generation of power and privilege, life-style and motive. Hence, in wrenching something called "the economy" from the larger "society," economics performs an operation that is intrinsically self-defeating. For the economic variables cannot be excised from the larger social system in which they are embedded and treated as a microcosm of that system without seriously distorting the relation of the model to the reality.

For example, economic analysis is frequently employed to explore the "growth paths" over which the economy—and of course the surrounding society—might travel. But the trouble with the analysis of these growth paths is that they do not take into consideration the social changes or the political frictions that growth will generate. The models assume that everything "noneconomic" will go on as it is, so that the only important changes will be the quantifiable relationships of capital and output and savings, etc. Yet it is apparent that even a single generation of growth at the rate of the past will bring average household incomes into the "full professor" range of $18,000 (in terms of today's purchasing power). Such an upward shift in the socioeconomic center of gravity is virtually certain to affect patterns of incentive and activity, particularly among the lower income groups, and might severely change the distribution of power among the various groups competing for shares of the national income. Yet, before these potential developments, economics is silent, *although these repercussions are likely profoundly to affect the validity of the growth projections themselves.*

There is, of course, a reason for this restricted scope of economic analysis. It is that neither economists nor anyone else are in a position to state with any degree of certainty what the social and political effects of economic change will be. The economist is not heedless as to the wider implications of the economic model he manipulates. It is rather that he feels that he cannot say anything "scientific" about them, within the rules and conventions of the paradigm in which he works.

Here is where the "scientific" model of economics thus meets its ultimate limit. For in its unwillingness to indulge in sociopolitical "guesswork," out of fear of transgressing the limits of proper scientific procedure, economics is placed in a position in which its findings *must* be partial and incomplete and very possibly erroneous. For every economic act, from the simplest response to a change in prices to the most complex decisions of an entrepreneur, is also a social act. Every action taken by the actors in a social system is freighted with innumerable significances, which we clumsily categorize under the rubrics of "economics" and "sociology" and "politics," etc. Here is the problem of abstraction that we encountered as the initial stumbling block in the approach to economics, come home with a vengeance at a much deeper level of significance. No one denies that abstraction is an essential precondition for a social science if it is to reduce the complexity of the real world to manageable proportions. But we can now see that the sharper and clearer the abstract model we create, the less "interdisciplinary" that model must be—*and insofar as the reality of the social process is unitary and indivisible, the less reliable as a guide to the social mechanism it must also be.*

Must economics resign itself to the "irrelevance" of an avowedly inadequate grasp of society? Or should economists abandon the scientific paradigm that has given such distinction to their efforts?

There are, so far as I can see, two ways out of the impasse. The first is to seek to widen the existing model of society by deliberately setting out to add as many social and political linkages as possible to the economic armature. The problem is how to go about this ambitious task. We stand today before the ramifications of social action much as did Linnaeus before the disorganized array of living forms, and we seek, as he did, a principle for introducing order into seeming chaos. Because we do not have that organizing principle, there is little we can do at the moment except to search for one, and that will necessarily impose a considerable burden on economic thought. For the search requires that we bring into sharp focus the hitherto overlooked noneconomic aspects of the familiar economic variables. The economist, in building his models, will have to learn to think in units much richer—but correspondingly much less easily manageable—than heretofore. His one-dimensional "consumers" must now simultaneously assert their existence as voters, members of social groups, role-actors under various forms of stress, etc.; his "firms" must figure as *loci* of political activity and of industrial discipline as well as of production. The state, technological change, economic growth—in short, all the variables that are now disposed of as letters in the equations by which the processes of economic society are described—will have to "mean" something quite different from what they do now.

It is enough merely to suggest such an enlargement of the model to recognize its difficulties. For we are asking for an extension of the orderly patterns of the scientific paradigm to cover a much wider expanse of life than we have hitherto been able to encompass within the bounds of equations. Indeed, it may be that we are asking for an extension of the paradigm beyond the limits to which it can properly be applied. Undoubtedly there are *some* linkages in the social process that patient investigation will enable us to incorporate into a richer model of society, but it seems very probable that many others will remain elusive, or indeterminate (as we have seen economic behavior itself to be under certain circumstances), or simply too complex. In the end, the very ideal of a scientific model capable of representing social change may be beyond hope of attainment.

In that case where can economics go? There is one direction, I would suggest, that would enable it to retain the marvellous architecture of the scientific model, while shedding the restrictions that rob that model of so much relevance. *This would be the conversion of economics into an instrument of social science whose purpose and justifi-*

cation was not so much the elucidation of the way society actually behaves, as the formulation of the ways in which it should behave. To put it differently, it would change the purpose of economics away from the discovery of the consequences of presumably known behavioral tendencies to the specification of the necessary behavioral patterns to enable society to reach a postulated goal.

This reorientation of economics, which has been called "instrumental" by its originator, Adolph Lowe,[5] would retain the "rigorous" aspect of the scientific method, but would subordinate its "value-free" aspect to an explicit recognition of the subservience of economic techniques to political and social ends. In the instrumental use of economics, the initial task—into which the economist would enter only as a member of the polity, quite without any special claims to expertise or superior wisdom—would be the articulation of desired social destinations or goals. These might be quantitative or qualitative, reformist or revolutionary, and they would be justifiable only by whatever standards of morality prevailed in society at that time. (To the horrified exclamations over such a meddling in human affairs, the instrumentalist answer is that the passive stance of economists today in fact constitutes an affirmative vote for whatever goals are inherent in the given socioeconomic mechanism at the moment.) The first task of an instrumental economics would thus be to raise the problem of social destination to its proper level as the most important decision that must be made by any society.

Once a social destination has been posited (if only as the premises of an economist's model), it then becomes possible to apply the "scientific" procedures of economic analysis, supplemented to whatever extent we can with sociological and political linkages. For the task of the economist—who is now properly on the way to becoming a true social scientist—is to describe the various ways in which a social system that finds itself at point A may begin to move to point B. This description must include not only the necessary technical steps, such as the amassing or reallocation of capital or other resources, but also the sufficient behavioral steps, including the political and social as well as economic actions that will be compatible

[5] Adolph Lowe, *On Economic Knowledge* (N.Y.: Harper & Row, 1970), as well as *Economic Means and Social Ends*, R. Heilbroner, ed. (N.Y.: Prentice Hall, 1969).

with the final target. And in turn these behavioral conditions will cause the social scientist to ask what institutional settings, what political directives, etc., may be needed as a still deeper-lying prerequisite for the desired social change.

It is enough merely to specify the width and depth of the necessary instrumental analysis to indicate that such a redirection of economics would be very nearly as challenging as the extension of the scientific model we previously discussed. The difference, however, is that a failure to specify all the various interactions of the social process does not vitiate the instrumental model, as it does the conventional "predictive" model. It simply means that various policies may have to be tried on a pragmatic basis to see how well they succeed in bringing about the desired result. Instrumentalism thus does not require a set of universal laws of behavior, but only a commitment to the conscious subordination of behavior to the requirements of deliberately selected social goals. It is, in Walter Lippmann's apt phrase of many years past, the choice of mastery over drift, with all the dangers and difficulties that any such assumption of control must bring.[6] The justification, if in fact one is needed, is that the social universe, unlike the natural universe, is not a mere collection of objects doomed to obey the laws of nature, but an assembly of human beings who include among their attributes the possibility of affecting the relationships that make up the system of which they are a part. Economics can thus find its ultimate relevance by becoming the instrument through which men can, within the limits of their knowledge and wisdom, use the procedures of science to gain those ends they hold in high value.

---

[6] It may be, of course, that the social goals include freedom of behavior—i.e., the decision not to intervene into the decision-making process. In that case, it may be impossible to achieve other goals that would require the influencing of behavior. In such a case, instrumentalism serves the purpose of making clear what opportunity costs must be borne in opting for behavioral freedom.

CAPITALISM AND THE ECONOMISTS

5

Science and ideology in economics

ROBERT M. SOLOW

THESE notes are intended only incidentally as a commentary on Robert Heilbroner's article. I take up some of the same questions and give independent answers, sometimes similar to his, sometimes different. The questions confront any teaching economist these days who talks to his students and reads the handwriting on the wall.

Relevance to what?

Utility theory is a prize example of a certain kind of economic theory. It sets out to be an abstract theory of rational choice, free of deep psychology, free of social institutions, applicable equally under capitalism and socialism. Naturally, it turns Freudians purple and sociologists green. Heilbroner's student wants to know if the theory of diminishing marginal utility is relevant. In Perry Mason stories, the judge is supposed to rule on the relevance of some line of questioning *to* something very particular, namely the case being tried. What does the student have in mind? Is diminishing marginal utility relevant to "ghetto life or the behavior of corporations or the military-industrial complex"? I have a short answer to that question: Not very. If the student is interested only in ghetto life, the behavior of corporations and the military-industrial complex, then he can probably survive without knowing much about diminishing marginal utility (though it might actually help with the behavior of some corporations). I have never promised that every bit of economics he learns will be relevant to those three things.

If, on the other hand, Professor Heilbroner's student is interested in the proper scale of fares on public transportation and congestion charges on urban roads, in the hope of making cities more livable, or if he is interested in changing the income-tax schedule so that it will distribute incomes more equally without turning off valuable effort from talented people, or if he wants to think about the way riskiness affects a man's choices of securities for his portfolio (or a government's choices of flood-control projects), then he might do well to master the idea of diminishing marginal utility. I couldn't care less either way, though naturally I expect every student of economics to learn about the foundations of the subject, even if bits of what he learns are irrelevant to his special interests. Do medical students complain that they have to learn about livers when they are specially interested in ear-nose-throat?

There is, however, a different point to be made here. Since the 1930's and until recently (perhaps even now), much of the intellectual energy of the profession has gone into macroeconomics, the analysis of the main global totals that describe the functioning of the economy as a whole: the national product, the unemployment rate, the price level, interest rates, etc. That was certainly relevant to the understanding and elimination of mass unemployment, which was probably the main domestic social problem of the 1930's. In fact, even unemployment of the magnitude of 7½ per cent of the labor force, as recently as the summer of 1958, was nothing to sneeze at. I don't think economists need apologize for having worked hard on macroeconomics into the 1960's. Even now, the combination of 5 per cent unemployment and rising prices poses a mean difficulty for policy. I, personally, think the current inflation has been inflated as a social problem, but if fear of it leads to policies which drive unemployment up to 5 or 6 per cent (and 10 or 12 per cent in the ghetto), then perhaps macroeconomics still has a bit of relevance.

Many students—and others—are not turned on by macroeconomics because their feelings about social priorities have changed and are changing. The macroeconomic problems now seem (and no doubt are) less urgent than the problems of war, poverty, racial discrimination, urban decay, traffic congestion, and the power of the large corporation. That's where the relevance is now. Economics tends to respond to shifts of opinion like this, but not overnight. It takes time for middle-aged men to change their research interests and their teaching, and it may take even longer for them to drum up any interesting and useful things to say. The theoretical analysis may be difficult, and statistical data are rarely available about something that has just now reared up in public consciousness.

After the war, for instance, the economic development of Asia, Africa, and Latin America became an interesting and important problem, became suddenly relevant, that is, for the obvious reasons. There were very few answers to begin with. In time, there came into being a corps of talented and specialized people whose main interest is the study of, and promotion of, economic development. It is a difficult matter, in which economic forces and other forces interact closely; the data are often scarce and unreliable; analogies to what goes on in developed economies may break down; powerful interests pull in different ways. Twenty years later, I'm not sure there are many answers yet, but not for want of looking.

There is already a shift of activity underway in the direction of work on the sort of problems that now seem most urgent. On some of the easier ones, like traffic congestion, there is already a certain amount of knowledge. Others will take more time and more work. (And, by the way, the best allocation of scarce research resources among competing ends is a good exercise in the theory of diminishing marginal utility.)

Are economists conservative?

Professor Heilbroner thinks that economists stay away from certain subjects because they are dangerous. To study them is almost automatically to be critical of the System, or at least of the System's pretensions. Because economists are generally conservative, in the sense that they are not disposed to entertain any radical critique of the foundations of the System, naturally they shy away from exposing the seamy side of things. According to George Stigler, economists are generally conservative because the scientific study of economics makes them so. According to Professor Heilbroner, the more likely explanation is that economists tend to take care of the System because the System takes care of them, by paying professors' salaries well into the upper reaches of the income distribution.

I have my doubts about that last bit of reasoning. It seals off discussion. If I disagree on an issue, the implication is that I am a paid lackey of the System. If I protest that I, like Professor Heilbroner, am above that, I am doubly suspect. It is like what happens if I say that Freudian theory is obvious nonsense: I am told that I only say that because of my relation with my mother. I protest that my mother had nothing to do with it. "See!" says my Freudian friend and walks away a sure winner. (Even in the days when I was a close student of Marxism I used to wonder about a similar question: if social theory is part of the ideological superstructure, hence not to be taken at face value, why is not Marxism also part of the superstructure—and in that case why should we take at face value the Marxist notion that social theory is part of the ideological superstructure?) Moreover, I rather suspect that many more radicals are to be found among those, and among the children of those, with incomes near $21,000 than among those with incomes nearer the median. Still, Heilbroner does well to be suspicious.

But I really want to comment on Stigler's proposition, which Heilbroner accepts. I am interested in it because I have been exposed to the scientific study of economics and do not regard myself as politically conservative; and also because it seems to me that economics has quite a strong radical potential too, especially along equalitarian lines.

Are economists really conservative? As it happens, there is now some evidence on that. Recently the Carnegie Foundation collected questionnaire responses from a large sample of academic people. Respondents were asked, among other things, how they would characterize themselves politically among the possibilities of Left, Liberal, Middle of the Road, Moderately Conservative, and Strongly Conservative. Of the economists, 61.7 per cent classified themselves as Left or Liberal, as compared with 47 per cent of the whole sample. The economists were more conservative than the sociologists (80.5 per cent Left or Liberal), political scientists (71.8 per cent), psychologists (69.1), anthropologists (69.4), historians (68.7) and teachers of English (65.8). They were less conservative than mathematicians (47.3), chemists (44.8), physicists (54.4), biologists (44.90), foreign-language teachers (56.0), medical school people (40.6), teachers of education (44.7) and of engineering (28.9).

One might be led to the conclusion that the study of economics makes one more conservative than does the study of sociology or political science, but makes one more radical than does the study of biology or mathematics or chemistry. I suspect that's not a good conclusion either. My guess is that it will be found, for example, that there is a fair correlation of place in the political spectrum with age, and that economists are on the average a bit older than sociologists and perhaps political scientists. (I have no idea about the natural scientists.) There may be other such factors at work. And it may well be that the causal arrow runs from political disposition to choice of field as much as the other way. In any case, it seems to be untrue that economists are particularly conservative, or that the study of economics makes them so.

It may be the case that some radicals unimproved by the study of economics tend to accept certain soft-hearted shibboleths that will not stand up under cold analysis. It may equally be the case that certain conservatives laboring under the same handicap tend to accept certain hard-hearted shibboleths that fare no better: such as that the free market always leads to better results than would come from intervention, or that public enterprise is inherently wasteful, or that steeply progressive taxation is unjust and inefficient, or that the AFL-CIO runs the country.

But then why do economists shy away from those dangerous questions? I think there are different reasons for different questions. Most economists would regard the political and community behavior of corporations as outside their field of competence. Maybe they're too modest. A study of "the size and distribution of the benefits of the war economy by socioeconomic grouping" scares me more by its impossible difficulty than by its possible subversiveness. I doubt that the data are available to do such a study with the rigor and precision the profession now demands. And beyond that, who is to know what Congress would otherwise have done with the resources that it actually voted to the military? (By the way, for an example of what can be done, see "Fiscal Policy and the Poor: The Case of Vietnam," by Charles E. Metcalf, in the Winter 1970 issue of the Kennedy School journal *Public*

Policy.) There actually have been attempts to analyze the "benefits accruing from 'public goods' by income strata"; it is indeed a fairly standard exercise, though the results are partly conjectural because simplified rule of thumb is all that is available for working out the incidence of some public expenditures. In the case of United States investment abroad, I don't feel competent to guess at the true state of affairs. I do not mean to suggest that economists are not as much afflicted with blandness as others, or that some may prefer, consciously or unconsciously, to avoid touchy subjects. I do wonder why those touchy questions one is accused of avoiding are so often vague and unanswerable questions.

Value-free social science?

It is sometimes said that academic economics and other social sciences are necessarily ideological, that their alleged "objectivity" is at best naive and more likely fraudulent. The claim to scientific objectively is a swindle; it permits ideology to masquerade as science.

No doubt some research is slanted; the results are decided before the data are in, or the data are carefully selected to prove a point. One hopes that professional criticism will catch this sort of thing but, inevitably, some of it escapes exposure. Where powerful interests are at stake, some research will be consciously or unconsciously perverted, and the critical mechanism will be diverted or dulled. But that is not what we are talking about now. Something subtler and deeper is supposed to be the case. Social scientists, like everyone else, have class interests, ideological commitments, and values of all kinds. But all social science research, unlike research on the strength of materials or the structure of the hemoglobin molecule, lies very close to the content of those ideologies, interests, and values. Whether the social scientist wills it or knows it, perhaps even if he fights it, his choice of research problem, the questions he asks, the questions he doesn't ask, his analytical framework, the very words he uses, are all likely to be, in some measure, a reflection of his interests, ideologies and values.

It is important to keep these different kinds of biases distinct. Some are more important than others; and, without care, the existence of one kind may be mistakenly thought to imply something about another kind. It is undeniable, for example, that many of the economist's terms, however technically and naturally they are defined, have definite overtones of value. "Equilibrium" is an obvious case in point: "market imperfection" is another. These are, I must stress, technical terms: their definitions are not value-loaded. On the other hand, it may be, as they say, no accident that "equilibrium" sounds good and "market imperfections" are obviously far from perfect. This sort of bias is not really very important. It is also correctible, but rather little follows from that. Suppose "equilibrium" were replaced by a more neutral phrase like "state of rest" or, if that is too evocative, by the mathematical term "singular point." Does anyone think that bourgeois society would totter? Besides, if the concept itself is naturally value-loaded, then I fear that after a few decades people would come to regard a "singular point" as a big deal. But evidently much more than this is at stake.

Here is Gunnar Myrdal, for instance: ". . . no social science or particular branch of social research can ever be 'neutral' or simply 'factual,' indeed not 'objective' in the traditional meaning of these terms. Research is always and by logical necessity based on moral and political valuations, and the researcher should be obliged to account for them explicitly."[1] Myrdal does not mean only that no amount of "objective" collection and analysis of data, no amount of "scientific" theorizing could ever eventuate in a policy recommendation without some reference to a scale of values. That is true, but would hardly be news. It is the orthodox view of the theory of economic policy. But Myrdal speaks of "false claims of being able to ascertain relevant and significant facts . . . without explicit value premises." On the other hand, he is curiously inexplicit about the precise way in which open or concealed valuations manage to shape and color even the simplest factual and analytic research projects. It is a little hard to see how ideology sneaks into an attempt to discover how purchases of frozen orange juice respond to changes in price (even a socialist planning board might want to know that), or—to take something more specifically capitalist—how the plant and equipment spending of corporations is related to their sales and profits, interest rates, stock prices, taxes, and other things. Perhaps the idea is that the demand for orange juice would behave differently in a society with a radically different distribution of income, and certainly the

[1] *Objectivity in Social Research* (NY: Pantheon, 1969), page 74.

investment study is tied to the institutions of private property. But neither does it seem in principle very ideological to study the working of a capitalist economy, if that is the kind of economy we have.

Sometimes Myrdal says things that seem to contradict his major line. He says at one point: "The relative importance of nature and nurture is a question of facts, and beliefs can be proved to be true or false by research." I don't see how this is consistent with the impossibility of objective social science. Indeed, a statement like the following is capable of more than one interpretation. "When these valuations have been brought out into the open, anyone who finds a particular piece of research to have been founded on what he considers wrong valuations can challenge it on that ground. He is also invited to remake the study and remodel its findings by substituting another, different, set of value premises for the one utilized." The difficulty here is to know what it means to "remodel" the findings of research. If it means that the same research results will lead to different practical conclusions with different value premises, we are back to what everyone knows. If it means that somehow the research results will be different, then it is not clear how or why.

The whole discussion of value-free social science suffers from being conducted in qualitative instead of quantitative terms. Many people seem to have rushed from the claim that no social science can be perfectly value-free to the conclusion that therefore anything goes. It is as if we were to discover that it is impossible to render an operating-room perfectly sterile and conclude that therefore one might as well do surgery in a sewer. There is probably more ideology in social science than mandarins like to admit. Crass propaganda is easy to spot, but a subtle failure to imagine that institutions, and therefore behavior, could be other than they are may easily pass unnoticed. It may even be that perfectly value-free social science is impossible, though I regard that claim as unproven about the kind of work that has genuine claims to be science. But suppose it is so. The proper response, I should think, would be to seek ways to make social science as nearly value-free as it is possible to be (and, of course, to be honest about the residue). The natural device for squeezing as much unacknowledged ideology as possible out of the subject is open professional criticism. Obviously, then, one must protect and encourage radical critics. I think that outsiders underrate the powerful discipline in favor of intellectual honesty that comes from the fact that there is a big professional payoff to anyone who conclusively shoots down a mandarin's ideas.

Interdisciplinary economics?

When you leave your car with an auto mechanic, it doesn't bother you that he will regard it just as an internal combustion engine on wheels. You don't feel it necessary to remind him that it is also a status symbol, an object of taxation, and a possible place to make love. Why, then, is it bound to be wrong for economists to regard the economic system just as a mechanism for allocating resources and distributing income, despite the fact that it also plays a role in the determination of status, power, and privilege? Why should economics be "interdisciplinary"? The answer is, presumably, because otherwise it will make mistakes; the neglect of all but the narrowly economic interactions will lead to false conclusions that could be avoided. The trouble is that the injunction to be interdisciplinary is usually delivered in general, not in particular; it is presented as self-evident, not as a conclusion from the failure of certain narrow undertakings and the success of certain broad ones.

I imagine that biochemistry and biophysics got started not because someone thought that biology should be interdisciplinary as a matter of principle, but because concrete research problems arose on the borderline of the biological and the chemical or physical. I will in a moment mention some problems that lie on the borderline between economics and the other social sciences. I think that the only way the interdisciplinary approach will ever make it will be for someone to make a killing on these or similar problems. But first there is a red herring to be cleared out of the way. The charge that economics limits itself by choice to the "quantifiable" is mistaken. There are plenty of qualitative distinctions that play a central role in economic analysis: employed-unemployed-out of the labor force, owner-renter, export commodity-import commodity, agriculture-industry. Sometimes, when people say that economics limits itself to the "quantifiable" they really mean to the precisely definable. When they want economics to be broader and more interdisciplinary, they seem to mean that they want it to give up its standards of rigor, precision, and reliance on systematic observation interpreted by theory, and to go over instead to some looser kind of discourse in which propositions are not supposed to be tested or testable, and it is never clear what kind of observation would definitely mark a hypothesis as false. One should resist any tendency for bad social science to drive out good. Mind you, economists talk a better game than they play. The fact that one can't experiment but only analyze history's single run of data means that an ingenious and determined man can keep a played-out theory alive for years. Nevertheless, it is a sound instinct to want to look at "the numbers," because they define, in a sense, what you're talking about. But any precisely defined measurement or observation would do as well.

There really are some problems outstanding that seem to call for an interdisciplinary approach. For example, five years ago the National Commission on Technology, Automation and Economic Progress tried to survey existing knowledge on job satisfaction and dissatisfaction under modern industrial conditions. There was pitifully little. With all the talk about alienation, dehumanization, and the loss of satisfaction from work, you would think that many researchers would be trying to find out the facts, by asking questions and by devising more direct measurements, by trying to figure out what aspects of particular routines are most destructive of satisfaction, and what loss of production would result from changing the routines. But apparently not so, unless the last five years have seen more such work than the period before.

Here is another example. Every discussion among economists of the relatively slow growth of the British economy compared with the Continental economies ends up in a blaze of amateur sociology. The difference is the bloody-mindedness of the English worker, the slowness of English management to adopt new products or new processes or new ideas, the elaborately amateur character of English business practice, the excessive variety of English goods corresponding to a finely stratified society, or the style of English education and the attitudes it imprints on graduates, or the difference is all of these in unspecified proportions. This may just be a complicated way to admit ignorance. More likely it suggests that the identifiable purely economic factors do not account for the full difference between the growth of productivity in Britain and in, say, Germany or Sweden. It is a fair, if very complicated, problem in social science to measure the other social forces that operate on the level of output per man and the growth of output per man, and perhaps to estimate the extent to which each of them is responsible for the British lag.

Finally, more in the spirit of the game, let me mention a much more vague sort of problem. It will, I fear, be more attractive to some, primarily because it is less capable of an answer. Never mind, here it is. One has the impression that wages and prices in the United States are rising a bit faster than they "ought to." More precisely, statistical analysis of the past gives us certain relations connecting the rate of wage inflation, say, to other things in the economy; given the current state of those other things, the rate of wage and price inflation may now be a little higher than the historical relation would predict.

By itself that would be nothing remarkable. Econometric relations have been known to break down before. Sometimes they never recover; sometimes the data creep back into consistency with the old relation and one must then try to decide whether the interlude has something systematic to teach or was simply the result of what is grandly called a random shock. What is interesting is that much the same thing seems to be happening simultaneously in other countries as well. The British are threatened with a wage explosion though unemployment is unusually high. The same seems to be true in Germany, with quite a different labor-market tradition. I am told, casually to be sure, that the same thing is happening in Sweden and France and probably elsewhere; they are all "off the curve" in the same direction. Why? Is it to do with rising standards of living, or with the attitudes of younger workers, or is it just a random shock?

Heilbroner is rather pessimistic about the prospects of a broadened interdisciplinary economics. So am I. If there is any hope for it at all it will come from solving well-defined, maybe even "quantifiable" problems, not from methodological precept.

Obviously, economists do recommend policies. These recommendations must rest on some ethical judgments or values. What are they? Sometimes they are rather "low-order" or specific value judgments with which no one would quarrel unless he were especially quarrelsome: unemployment is a bad thing; or inflation is a bad thing; it is better that people should have adequate food and housing than that they should not. Sometimes there is room for argument, or at least there is argument: if I say that unemployment is a worse thing than inflation (quoting some numbers, of course), someone may disagree. But at a deeper level, the theory of economic policy is supposed to rest on a "weak"—that is, widely acceptable—fundamental value premise: a change in economic arrangements is a good thing if it makes everyone better off, or at least no one worse off (counting the possibility that there are initially gainers and losers, but the gainers compensate the losers, so that in the end everyone gains). There are obvious technical problems with this criterion: for example, does "everyone" include foreigners and unborn generations? But they need not concern us here.

To get anywhere, there must be room in economic theory for tougher value judgments that weight the welfare of one person or group against the welfare of another. But the basic principle that makes market mechanisms attractive to economists is the one I have just stated.

What does it mean for someone to be "better off"? Many criteria are possible: you could say that a man is better off if he makes a bigger contribution to the health of the State or the glory of God. In economic theory, however, it has usually meant that he is better off *in his own estimation*. If you want to know whether A prefers working over a hot stove or in a nice cool sewer, you ask him; or better still, you offer him a choice between the two jobs and see which he chooses. Similarly for bundles of consumer goods. This criterion is clearly a certified product of nineteenth century liberal individualism, and in that sense economics is permeated with individualism.

This individualist orientation and the accompanying doctrine of "consumer sovereignty" are under attack these days, usually for the wrong reasons. It is said that ordinary people can not be entrusted with the judgment of their own welfare, not even with the choice of the things they buy. This is because they are ignorant of "true" satisfaction, or because they are manipulated by advertising, or because their tastes, such as they are, have been formed by a wicked society to preserve itself. We are all, or most of us, in the position of the Indian brave who has been hooked on cheap fire water by the greedy fur trader; in what sense is he better off because he willingly trades muskrat for rotgut?

No one can deny that advertisers advertise, and must have some effects (though one could argue about how much) on the preferences of consumers. It is certain that our preferences are far more social than biological or individual in origin. What should we conclude from these propositions? From the first, perhaps that advertising ought to be limited by taxation or regulated as to truth. From the second, what? Not, I hope, that individuals' judgments about their own welfare should not be respected, whatever their origin. One need only ask what could be put in their place—presumably the judgments of an elite. The attack on consumer sovereignty performs the same function as the doctrine of "repressive tolerance." If people do not want what I see so clearly they should want, it can only be that they don't know what they "really" want.

Now there are critical things that need be said about individualist welfare economics, but they are rather different things, well known inside the profession but not always understood outside it. For example, as I mentioned earlier, the individualist criterion is not complete. There are many questions on which it does not speak at all, and it *must* be supplemented by sharper value judgments that each society must make in its own characteristic way. For example, the strict individualist criterion does not imply that it would have been a good thing for the English government to have relieved starvation in Ireland during the potato famine. It was, however, a fallacy for the English to believe, as some of them did, that the criterion did imply that it would have been a bad thing. It is simply a separate decision. The individualist criterion does not say that the rich should be taxed to help the poor, nor that they shouldn't. It does say that if you decide to do so, it is in general better to give the poor money than to give them what you think they ought to consume. It takes a bit more than diminishing marginal utility to imply that you should in fact redistribute income toward the poor, but it sure helps, as Professor Heilbroner remarked.

This is not a trivial matter. I said that the choices must be made, and default is merely a way of making them. The radical critique is right that merely to mumble something about not interfering with the market is to favor the current holders of wealth and power.

Another problem arises not so much with the individualist welfare criterion itself as with its implementation through the market. The market mechanism works best when it deals with "private goods" which have to be divided up among individuals or families and consumed by them separately. When "consumption externalities" are important, when one person's consumption decisions have a strong effect on the satisfactions of others, or where different people consume the *same* "public good," like clean air or weather forecasts, the market need not work so well, and other, more directly political, decision methods may take its place. The trouble is that each producer and consumer compares the market price with benefits to himself; other benefits and costs are not taken into anyone's account. This is also a serious matter. As standards of living rise, population density increases, and technological interactions grow more pervasive, it may be that a greater and greater part of economic life will have to come under these rules of the game, which may turn out to be quite different from the rules of the private-property game.

But notice that neither of these problems I have sketched involves superseding the individual as the best judge of his own satisfactions.

Is science necessary?

The modern critics of economics and the other social sciences rarely seem to do any research themselves. One has the impression that they don't believe in it, that the real object of their dislike is the idea of science itself, especially, but perhaps not only, social science. A sympathetic description of their point of view might be: if the ethos of objective science has led us to where we are, things can hardly be worse if we give it up. A more impatient version might be: what good is research to someone who already knows? The critics, whether from the New Left or elsewhere, do not criticize on the basis of some new discovery of their own, but on the basis that there is nothing worth discovering—or rather that anything that is discovered is likely to interfere with their own prescriptions for the good society. My own opinion is that the good society is going to need all the help it can get, in fact more than most. A society that wants to be humane, even at the cost of efficiency, should be looking for clever, unhurtful, practical knowledge.

Even Professor Heilbroner, who is a reasonable man, speaks of the "scientific" paradigm, in quotes, as if there is some other way to find out about the world. There is no other way. Economics can be better or worse science, but it has no other choice. Breadth is not the issue either. What Thomas S. Kuhn remarked about natural science holds for social science too: ". . . though the scientist's concern with nature may be global in its extent, the problems on which he works must be problems of detail." I do not fully understand Heilbroner's "instrumental" road to relevance, but so far as I do understand it, it seems likely to make more, not fewer, demands on "positive" economics. It is hard enough to understand the working of the institutions we have; to understand the working of hypothetical institutions may sound easier, but only because our mistakes would be less likely to be found out.

My argument about science is not a defense of what is sometimes called "scientism." The possibility of scientific economics (whether pure or merely as nearly pure as one can make it) does not entail the impossibility or the degradation of intellectual dissent and thoroughgoing criticism of the justice and quality of social and economic institutions. It seems reasonable, however, that whatever results can be established (after the most abrasive criticism) by economic science ought to set certain ground rules for philosophical and ideological discussion of economic institutions. The two kinds of activity ought to be distinguished, if not necessarily separated, simply because the reasons for believing one kind of statement are different from the reasons for believing the other. The main trouble with the attack on "scientism" of someone like J. P. Nettl[2] is that it fails to come to terms with genuine, limited, reasonably "objective" social science. The free-floating intellectual's social imagination is not completely free to float.

It makes no sense to squabble over which is the "higher" and which is the "lower" form of intellectual effort. But there is sense in a determined effort to see both that issues of value-conflict do not get smothered in smooth pseudo-science and that questions susceptible of a scientific answer not get submerged in a flood of ideology impervious to analysis and evidence.

---

[2] J. P. Nettl, "Ideas, Intellectuals, and Structures of Dissent," *On Intellectuals*, edited by Philip Rieff (NY: Doubleday, 1969).

STRUCTURAL CHANGES IN THE ECONOMY

6

The transformation of Wall Street

DANIEL SELIGMAN

LOOKING back at Wall Street in the 1960's, historians of finance will almost certainly conclude that the decade's most important event was the "institutional transformation"—no simpler term seems available—of the stock market. When the 1960's began, pension funds, mutual funds, insurance companies and other institutions together held about 15 per cent of American-owned common stock. When the decade ended, institutions' share-holdings were up to about 25 per cent of the total; Manuel F. Cohen, the former chairman of the Securities and Exchange Commission, estimated the value of their holdings early this year at over $260 billion. In 1960, institutions accounted for less than 25 per cent of New York Stock Exchange volume. The figure now is over 40 per cent; in addition, a significant and rising amount of institutional trading now takes place off the board.

The larger role of the institutions is by now a familiar proposition on Wall Street and has been noted in the press. Somewhat less attention has been paid to the broad exodus of individual investors from the stock market. The fact that they are getting out might be inferred from the corollary fact that the institutions have been going in. But the business pages have for some reason never really focused on the persistent and heavy net selling of stock by individuals.

Before the 1950's, any such phenomenon would have been impossible because the institutions that bought stock were few and relatively small; every time an individual sold stock, there was almost certainly another individual on the "buy" side of the transaction. During the 1950's, the growth of pension funds created a large potential buyer for stock, and toward the end of that decade, these funds began to show a sizable appetite for stock. Still, until 1959 there had never been a year in which individuals were net sellers of stock.

Since that year, however, they have sold more stock than they bought in every year and in every kind of stock-market environment; and the rate at which they were selling seemed to be rising at the end of the decade. Indeed, one might sum up the market of the 1960's by noting that it consisted of individuals selling stock to institutions. In 1969, for example, individuals sold about $10 billion of stock on balance (the figure excludes mutual-fund shares). In the same year, private pension funds bought $5.4 billion worth of stock, public pension funds bought another $1.8 billion, mutual funds bought $2.5 billion, life insurance companies bought $1.6 billion, and other institutions bought several billion more.

The reasons why

There seem to be two principal reasons for this large and continuing transfer of stock from individuals to institutions. One reason, apparently, is that stock is worth more to institutions—first, because institutions are more likely than individuals to diversify; second, because it is easier for them to hold for the long term. It is clear that both differences critically affect the rates of return on stock investments. The massive study of stock prices in 1926-65, conducted jointly by the University of Chicago and Merrill Lynch, Pierce, Fenner & Smith, reached these conclusions, among others: 1) For investors who picked "small groups of stock at random, instead of just one, the risk of loss would have been considerably reduced and the probability of a larger profit considerably improved." 2) For investors not forced to sell during recessions, the "chance of making a profit and the amount of profit would have both been significantly increased." Thus it is reasonable to conclude that, in general, institutions enjoy higher rates of return than individuals do. And so it is also reasonable to assume that they have been bidding stock away from individuals.

Another reason for the institutionalization of the stock market may lie in the transformed economics of the stock-brokerage business. Because the business is labor-intensive, the cost of processing individuals' trades soared during the inflationary latter 1960's. During this same period, volume rose sharply: from less than 5 million shares on an average day on the NYSE in 1964 to almost 13 million in late 1968. In an effort to handle the accompanying paper blizzard, many firms both staffed up heavily and expanded their investments in data-processing equipment. Both changes had the effect of raising break-even points considerably. In 1964 a typical medium-sized broker with predominantly individual customers was probably in the black on days when volume ran over 4 million shares. By the spring of 1970, the broker was probably in the red on days when volume was under 12 million.

And by that time volume *was* under 12 million on most days. By that time, furthermore, many brokers could not operate profitably even on high volume if it represented an accumulation of small trades. The "Aunt Jane" who had been wooed so persistently during Keith Funston's years as president of the Big Board was now a most unappealing business proposition. Early in 1970 there was probably not a brokerage firm in New York that could make money by executing an order to buy 50 shares of a $20 stock: quite a few firms were losing money even on 100 shares. Everyone, however, had been able all along to make money on the 10,000-share blocks generated in institutional trading; and everyone, accordingly, had long since set up an institutional department. When Aunt Jane walked in the door with $1,000, her broker tried to talk her out of that $20 stock and to sell her a mutual fund. The broker's commission on the fund was considerably higher (the firm would typically receive $70 on a $1,000 purchase, compared to only $15 for an odd-lot $1,000 stock purchase); besides, mutual-fund owners do not trade as much, and so the probability that Aunt Jane would be back soon with another profitless proposition was reduced. Even Merrill Lynch, the world's largest and most profitable brokerage house, which had resisted mutual funds, regarding them as a threat to the basic brokerage business, stopped resisting in 1969 and began offering funds to its customers.

In short, one considerable reason for the sizable move of individuals out of stocks and into mutual funds in recent years is that their brokers have steered them to the funds. In 1969, when individuals were net sellers of almost $10 billion of corporate stock, they were net buyers of $5.6 billion of mutual-fund shares.

As the outlines of the new institutional market for stocks first came into focus a decade ago, Wall Street greeted it with immense satisfaction. To members of the New York and American stock exchanges, the institutions looked wonderfully profitable: revenues rose directly in proportion to the size of the big new orders generated by the institutions, while the costs associated with executing the orders were scarcely any higher than those on a 100-share lot. In addition, there was a widespread notion in the early 1960's that this large new institutional demand would act as a permanent support under stock prices. That is, there was clearly a large new participant on the demand side of the price equation; there was no reason to suppose that the supply of stock would be growing at a greater rate than it had; and so, the conventional wisdom of the Street had it, prices would be forced up. And that was not all: it was also part of the conventional wisdom that price changes would be less erratic as more of the stock came into institutional hands. Small investors might panic in declining markets and might be overenthusiastic in rising markets, in both cases exaggerating stock price movements. Institutions had better research and greater cash resources than individuals and would be able to prevent price gyrations from getting out of hand.

Every one of these assumptions is now in question. The brokerage community, though heavily dependent on institutional business for what profits there are today, is beginning to realize that the institutions pose a threat to the basic arrangements of the brokerage business and the exchanges.

To begin with, the profitability of institutional business has been undermined or threatened in several different ways. First, the Securities and Exchange Commission began in 1966 to press for a "volume discount" in commission rates, i.e., it wanted proportionally lower rates for big transactions—and was unwilling to accept the New York Stock Exchange's proposal that the change be accompanied by proportionally *higher* rates for small transactions.

The SEC has not yet determined what rates it deems appropriate; but in December 1968, while it was still holding hearings on the issue, the NYSE unilaterally implemented a new rate schedule that included volume discounts and, Manuel Cohen has estimated, reduced commissions overall by some $150 million a year. In addition, the Antitrust Division of the Justice Department has raised the question whether there should be any minimum rates at all, for the first time raising the prospect of free competition in setting commissions. The Street today is gloomily persuaded that free competition would be murderous; its anxieties are focused especially on the institutions, whose bargaining power would be far greater than that of individuals.

The view that institutions represent a permanent support under stock prices pretty much vanished in the 1969-70 bear market. In the long run, the view probably has a limited validity: if stock is indeed worth more to long-term holders, and if institutions on balance hold for longer terms than individuals, then it is doubtless true that institutionalization means somewhat higher prices. But quite a bit more than this was claimed in the early 1960's. It was then often argued that institutions—pension funds especially—represented a quite new source of demand; that they could be counted on to grow prodigiously; and that, investing for the long term in a permanently inflationary economy, they had nowhere to go but stocks for the bulk of their investments. Proponents of this view often sounded as though there was no price-elasticity at all in the demand for stocks. Beyond that, they often seemed to forget that the growth rates of institutional demand were approaching certain limits.

A diminished appetite for stocks

In the 1950's and 1960's, pension funds' appetites for stock were growing for four different reasons. First, more and more large companies began, after the 1949 steel settlement, to adopt pension plans for their employees. Second, they had to fund not only the benefits attributable to present and future, but also to *past* employment; in other words, the employer's contributions are especially steep in the fund's early years, while the employer catches up on past-service funding. Third, the notion that pension funds should own stock, which seemed rather adventuresome in the early 1950's, caught on steadily, so that the proportion of their assets fund managers wanted to be in stocks grew fairly steadily. In 1950 private pension funds had $6.7 billion of financial assets, of which 16 per cent was in corporate stock. In 1960 they had $38.2 billion of financial assets, of which 43 per cent was in stock. In 1970 they had slightly over $100 billion of financial assets and perhaps two thirds of that is in stock. Today most big corporations already have pension plans; their plans have largely funded the past service of their employees; and their stock ratios are probably about as high as they would like them to be.

Thus the demand for stocks attributable to pension-fund growth is approaching the point at which only the fourth source of growth is left—the slow, long-term growth associated with an expanding labor force. In the short term, furthermore, pension funds may now sometimes act to depress stock prices. As long as the funds were in general working to increase their stock ratios, it was unlikely that they would ever get to be net sellers on balance; but now, with their ratios presumably close to their long-term objectives, their managers are freer to decide that bonds are a better buy than stocks, and to liquidate substantial amounts of pension-fund stock—something the market had never yet experienced but that looked quite plausible in the spring of 1970. (In a speech to the New York Society of Security Analysts last March, Henry Kaufman of Salomon Bros. & Hutzler, a major institutional brokerage firm, pointed out that bond yields, at 9 per cent for high-quality issues, had reached the "expected long-pull return on stocks, including dividends and price appreciation." Kaufman added that, furthermore, "the high return on bonds currently is a contractual payment into the future while the 9 per cent growth on stocks is [only] an extrapolation of past trends.")

The notion that institutionalization of the stock market means a smoother ride for stock prices is also seldom heard these days. It is true that institutions turn over their portfolios somewhat less than individuals, but the difference is less than it used to be; and it is abundantly clear that the mutual funds, whose turnover rates are higher than those of other institutions, have by themselves a capability for generating turbulence in the marketplace. Some mutual funds in the last year or two have turned over their portfolios at rates of around 100 per cent—i.e., rates equivalent to the creation of an entirely new portfolio every year. The rate for all mutual funds in 1969 was 45 per cent—more than twice the rate of the mid-1960's. A fair number of corporations now regard substantial mutual-fund holdings of their shares with more alarm than satisfaction. The earlier notion that such holdings were "in strong hands" and not vulnerable to dumping in market panics has been replaced by a feeling that such holdings are the most vulnerable of all.

The four markets

Thus from the viewpoint of the Wall Street community, the institutionalization of the stock market no longer looks like the bonanza that was envisioned in the early 1960's: it is not as profitable as anticipated, it will clearly not be the support under stock prices that some anticipated, and it will clearly not do much to prevent wild swings in prices. Indeed, the new view of many members of the exchange community is that the institutions look profoundly subversive these days, that they threaten many of the exchanges' own institutions with obsolescence. Perhaps most directly threatened are those in the cockpit of the exchanges, the specialists. Traditionally, they have been responsible for matching up buy and sell orders on the floor and for preserving orderly markets; and they have come to seem especially inadequate in an institutional era. But the problems for the exchanges extend beyond the specialists, and the very notion that the exchanges are *the* stock market has to some extent been eroded in recent years. There is a countervailing notion today to the effect that they are really just one of four different markets.

The second market, of course, is the over-the-counter market, in which the "continuous auction" of the New York, American, and other stock exchanges—an auction presided over by the specialists—is replaced by a process of negotiation. The over-the-counter market is nothing new, of course, and has traditionally provided the means of trading unlisted securities. What *is* new is the so-called "third market," in which securities listed on the NYSE are themselves traded over-the-counter, i.e., at prices negotiated by the buyer and seller. The purpose of this arrangement is to accommodate institutions, whose purchases or sales are often so large that they overwhelm the traditional exchange mechanism: they have to pay more or receive less than the market price in situations where their own orders constitute the bulk of the demand or supply for particular issues; in addition, the exchange's specialist, who is supposed to step in himself when no other buyer or seller can be found, often finds himself financially unable to handle institutions' orders. Third-market dealers, who had their own substantial inventories of listed securities available to prospective buyers, and who were able and willing themselves to buy substantial blocks, began springing up in the early 1960's to fill this gap in the exchange mechanism. While the third market accounts for perhaps 5 per cent of the volume in NYSE-listed stocks, it is responsible for 25 per cent of the volume in a few heavily traded institutional favorites.

Toward the end of the decade, some members of the exchange provided another alternative to the traditional system by offering "block positioning," an arrangement in which large and well-heeled firms offer to buy stock in amounts that specialists cannot handle and hold the stock in their own accounts until they can market it; such firms often lose money on the "spread" between prices they pay and the prices they can get for stock, but they may recover the loss, and more, in commission fees. In 1969 there were over 500 transactions involving more than 100,000 shares each; about a fifth of them were handled by the largest of the block-positioning houses, the NYSE member firm of Salomon Bros. & Hutzler. Since the block positioners operate within the confines of the Big Board, they are technically part of the primary market; but it is clear that they, too, are chipping away at the traditional specialist system.

Potentially the most revolutionary development of all is the fourth market, which consists simply of transactions between institutions, without any brokerage firms or exchange mechanisms participating at all. In principle, of course, any shareholders can buy stock, from other shareholders, or sell it to them, or simply swap, and thereby avoid brokerage charges. Individuals find ways to do business directly with one another occasionally, and some institutions have done so—there is, for example, a long history of swapping between the Ford and Rockefeller Foundations. But for the fourth market to evolve into something more than occasional deals between institutions that happen to hear fortuitously about one another's interests, rather special forms of communication are required. Some enterprising salesmen have made it their business to keep records of institutions' portfolios and to look for swaps that seem logical. The most ambitious fourth-market operation is one called Institutional Networks Corp., which has been trying for over a year now to sell a large number of institutions on a sophisticated computer system that would allow those joining the system to trade with one another anonymously, rapidly, and commission-free (although fees would be paid to the network).[1]

Redefining the central market

The exchanges, and especially the Big Board, have repeatedly assailed all such efforts as threats to the central market, contending that the new markets for listed securities could not function without the pricing mechanism provided by the exchanges but that this mechanism is increasingly impaired as more and more volume drifts away from the exchanges. The contention seems less plausible today than it did when it was first articulated a decade ago. The main difficulty with it is that so much volume has already drifted away, and so much other volume that is still reported in the NYSE totals is in fact bypassing the exchange's pricing mechanism, that it is no longer clear that the exchange really *is* the primary market.

The Big Board is thus increasingly vulnerable to attempts to redefine the central market, e.g., as in an aggressive presentation made last fall to the SEC by Weeden & Co., a major third-market house. "We are not opposed to the Central Market Place Concept," said Donald E. Weeden in his testimony. "We give it our wholehearted support. However, the NYSE version of the Central Market Place is truncated and self-serving. By building a wall around its own special privilege, the Exchange has kept out the capital, the inquiry, and the participation of other market makers, and restricted its own members from seeking better markets away. The result is that the NYSE is itself fracturing the Central Market Place." Weeden went on to propose "an up-to-date version of the Central Market Place, involving computers and electronic display panels . . . enhanced by a 'public' ticker tape capable of reporting trades in all market places, including all exchanges, on a real time basis."

---

[1] For a detailed description of this operation, see "Challenge to the Brokers," in *Fortune,* April 1969.

The proposal may or may not be implemented, but it seems fair to say, in mid-1970, that there is nothing the least bit visionary about it. The new technologies have made it possible to redefine our securities markets. And the new institutions have made it probable that this will happen.

STRUCTURAL CHANGES IN THE ECONOMY

7

The two faces of economic concentration

M. A. ADELMAN

LAST year, in a widely publicized speech, Attorney General John N. Mitchell professed alarm over the rapidly increasing concentration of American industry. In a less publicized article, Professor Pashigian of the University of Chicago suggested various methods of explaining the familiar fact that industrial concentration had been so stable for so long. Oddly enough, the Attorney General and the professor were not really contradicting each other; they were not referring to the same thing.

"Economic concentration" or "concentration of economic power" calls up a vision of a corporation, with billions in assets, thousands of employees, elaborate staff planning, etc., etc. It "dominates" its market and decides prices and outputs, either alone or in concert with one or two of its fellow giants. Such "concentration" and "domination," presumably, is the typical pattern, or at least "the wave of the future." The "grand sweep" of the 20th century is away from competition and into the new era of the big firm, etc. All this is a good story but not necessarily a true story. An effort to see and analyze what is actually happening in the world makes it apparent that "economic concentration," so far from being a simple, massive phenomenon, is actually a vague phrase applied to two different kinds of measurement of two different phenomena. One is that of bigness, the other of market concentration, but the relation between the two are quite complicated.

Concentration ratios

The statistics of *market concentration* exist because economists have tried to put numbers into their analysis of monopoly and competition. Theory and observation seem to prove that it does make a difference whether there are few or many firms in a market. The ultimate in fewness is monopoly, where a single firm can do what is best for the industry as a whole because it *is* the industry as a whole. It has the power to control output: to let only so much of production on to the market as will yield the greatest money profit, or perhaps the most quiet managerial life. Short of actual monopoly, the fewer the firms, the easier it is for them to collaborate, to align prices and production so as to travel much or all the way toward monopoly. The more firms, the harder is such collaboration, and the more are they forced willy-nilly to act independently. At the extreme of such competitiveness, each firm always seeks its own profit, neither trying to serve the group industrial interest nor expecting anyone else to do so. Of course, even a very large number of firms can be regimented into monopoly through detailed agreements tolerated or enforced by government. But if these cannot be made, there is no control of supply, and no market power. Output is higher, prices are lower, and resources better used. It is less clear that the *long-run* performance of a competitive market is also superior to one where a small group has market power and controls supply. I think it is, but discussion would take us too far afield.

All this seems simple enough. But it isn't. To begin with, trying to measure the manyness or fewness of firms in an industry is difficult. Companies are usually of such unequal size that a simple head-count is useless. Suppose one firm had 95 per cent of a market and 100 other firms had the other 5 per cent. There would be literally 101 sellers. But it would be nearer the truth to call this a one-firm industry. Both statements would be imprecise, but the first would be wildly wrong in its implications.

In the 1930's, the Census Bureau first tabulated something called the "concentration ratio": the sales of the largest four companies in each industry as a percentage of total sales of all companies in that industry. (The number four had no particular justification; it was only that Census regulations forbid separate publication of the sales of smaller numbers of companies. Fortunately, it turns out that it does not seem to make much difference if we use the largest eight or twenty; industry rankings are hardly changed.) Since 1947, the Census of Manufacturers has regularly collected and published concentration ratios for each of 450-odd manufacturing industries, further subdivided into about a thousand product classes.

Now, manufacturing is certainly the heart of what is usually called "the industrial economy." But it is well to keep in mind that manufacturing represents only about 35 per cent of the private economy; we have no such systematic data for mining, construction, wholesale and retail trade, services, or the "public utilities" of transportation, communication, and electric power. Thus, the very biggest (e.g., A.T.&T.) and the very smallest companies fall outside of the range of what is periodically canvassed and reported by the Census.

Moreover, the concentration ratio—the sales of the largest four as a per cent of the total—is no precise measure. A concentration ratio of 50 could mean that the single largest company had 49 per cent of the sales, while the three next largest had ⅓ per cent each. A concentration ratio of 50 could also mean that the largest four had respectively 13, 13, 12 and 12. True, some more refined measures have been devised; at least one of them, the so-called Herfindahl index, seems much superior, but it has not been applied widely.

Even if it were, it might not help much. A far more important limitation of the concentration ratio is the uncertain denominator, "total industry sales." The boundaries of the "industry" or "market" are often fuzzy. Any official definition is a rough approximation at best, and often has little to do with buyers and sellers whose offerings and demands interact to make a price. The basic idea of "a market" in economic analysis is: a discrete area of economic activity, with a clear-cut boundary between it and the rest of the economy. What is produced outside this area cannot easily and quickly be substituted for what is produced within. In the real world, however, such boundaries are never all that clear-cut. If much of a product used in this country is imported or exported, the American sellers and customers are part of a world-wide market, and figures on domestic activity are only a truncated fragment of a larger whole. (Large imports or exports do not necessarily mean that the concentration ratio overstates true market shares; the big American companies may account for some or even most of what is produced abroad. Hence the true world-wide concentration ratio could be greater or less than the false national ratio.) Inside the United States, there may be a substitution from similar or identical products which happen to be classified as belonging to another industry, or from facilities which can easily be converted. "Industries" are collections of similar plants which may produce a varied assortment of products. (Fortunately, there are separate tabulations of product totals regardless of what industries they happen to be produced in. The result is a valuable opportunity to cross-check and get a stereoscopic view.) It might at first seem that because of substitution across product lines or industry lines, concentration is usually overstated; as with foreign trade, however, it can work both ways—the "substitute" may itself be produced by a large manufacturer of the original. But one limitation of the data works always in a single direction; where products are sold in regional or local—i.e., "insulated"—markets, concentration ratios come out higher, sometimes much higher, than the national average.

Finally, a concentration ratio is a snapshot and gives no indication of the way an industry is going. In a concentrated market, but a growing one to which entry is relatively easy, rewards to those who grab the largest share may be so great that an attempt to hold the line on prices cannot succeed. Prices and outputs may be highly competitive. But if entry into this growing market is difficult (e.g., because of patents), there may be less competition than there seems.

The facts of market concentration

In any given instance, therefore, a concentration ratio may not mean anything. But what cannot be said of any individual in a group may be truthfully said of a group. A single concentration ratio tells us little about a given industry at a given time, but much may be learned from analysis of groups of industries over time, or among countries or regions. So the close attention to dusty detail is bound together with broad comparisons in time and space.

The most important contribution of the statistics is perhaps not in the detailed numbers but the way of looking at industries. We see some familiar terrain in a new light and the single stereotype of "big business" disintegrates into a more difficult but more interesting universe.

For example, the "third world" of the less-developed countries is almost surely the most highly concentrated and monopolized. Because incomes are low, most markets are extremely small, with room for few rivals. Capital, skilled labor and know-how are scarce, innovation is risky, and starting new enterprises is that much harder. Hence, vested interests are that much more safe from new entrants. Growth and development would entail the breakdown of a host of miniscule monopolies. But economic development is usually planned and controlled by government, which restricts imports to conserve foreign exchange, protects nationalized enterprise against either foreign or domestic rivals, etc.—all to the effect, if not always for the purpose, of narrowing the circle of possible rivals.

In such countries as Great Britain and France, companies tend to be much smaller than in the United States, but markets are very much smaller, so that in any given market, concentration tends to be higher than in the United States. (The European Economic Community is now providing much bigger markets, within which the rivals will be bigger but also more numerous and hence more competitive.) Japan is, as in so many other respects, in a class by itself. Concentration seems to run higher than in the United States, but the importance of foreign trade is so much greater that the conclusion is in doubt.

In the United States, one can study concentration over time. A comparison between manufacturing around 1900 and in 1947 seems to show a substantial drop. But given the unsystematic data for 1900, it seems better to make a sure bet on a milder proposition: concentration could not have increased. Between 1947 and 1966, individual industries or products often change, but the ups balance the downs, and there is no net total change in industrial concentration.

But as the stability becomes obvious, it also is less satisfying. Why should there *not* be a change up or down? If we could explain the past stability, and identify the causes thereof, we might make an educated guess at the future. But since we have not yet reached the first stage, we ought to resist the impulse to extrapolate and predict that what has been will be. Mark Twain said somewhere that the Mississippi River is an alluvial stream which often cuts its bank and has in the course of years shortened its total length—hence any fool could see, and some fools would say, that in a thousand years Minneapolis would be a suburb of New Orleans.

One plausible theory for the steady level of concentration is that the larger the industry, and the faster its growth, the better the odds that concentration decreases. The bigger market works in two opposed ways: there is more scope for economies of scale, and the companies therefore tend to be bigger; but there is also more room for rivals, and the latter tendency seems on the whole to be a little stronger. For example, in 1963 there were 52 product classes with sales of more than a billion dollars each. Three of these products had a concentration ratio of over 80 (the largest four made 80 per cent of the sales) and 16 of them had a concentration ratio of over 50. At the other extreme, there were 96 product classes with sales of less than $50 million apiece. Fourteen had concentration ratios of over 80, and 47 of over 50. Thus, the bigger the industry and the bigger the companies, the lower the concentration. But the tendency is very mild.

The stereotype that the big company has a big market share is obviously supported by many examples. Only, it is refuted by even more. The association of rapid growth with lower concentration is somewhat stronger, but even so it is not very marked.

Yet if size of market and economic growth are on the whole favorable to lower concentration, one would expect a slow downdrift over time; in fact, there has been none since 1947. Perhaps the reason is that in earlier years the growing markets increased company size faster than market size. That, of course, is the problem with which we started! As in most fields of study, the more we learn, the more questions we encounter.

Attempts have also been made to relate concentration to other economic variables. The more concentrated an industry, for instance, the less competitive it presumably is, and the more profitable it ought to be. And so it appears in fact. But the relation is a weak one at best, and the odds are only a little over 50-50 that a more concentrated industry will be more profitable than a less concentrated one. This mild signal is actually reassuring. For if higher industry concentration were associated with *lower* profits, one would have to ask—on the assumption that businessmen are not as a class insane—whether concentration data made much sense.

The largest 100 - 200 - etc.

These facts on industry concentration get no headlines, and they are not what Mr. Mitchell professed to view with alarm. What we mostly hear about is "aggregate concentration" (or "overall concentration" or "superconcentration"), which is the percentage of total manufacturing—all industries added together—that is accounted for by the largest 50 or 100 or 200 firms, regardless of industry.

A few comparisons over time are possible. In 1935, the largest 50 companies accounted for somewhere around 22 per cent of total Census manufacturing "value-added" (i.e., the margin between purchases and sales); in 1947, only 17 per cent; in 1954, 23 per cent, and in 1966, 25 per cent. Another way to measure big-company participation is by the proportions of corporate assets owned. Assets are harder to measure and more risky to compare, because of variations in accounting rules, which moreover have changed over time. (In addition, it is necessary to splice together various statistical series.) But the figures on assets show a parallel movement to the figures on "value-added." According to the statistics of the Internal Revenue Service, the largest 139 manufacturing corporations had 47 per cent of all corporate assets in 1931 and the largest 141 had 48 per cent in 1963. (Technical adjustments, for noncorporate manufacturing and for greater consolidation of corporate reporting, work both ways and make little net difference.) Measuring on a different basis, Professor Charles Berry has estimated the share of the largest 100 at 44 per cent in 1948 and 48 per cent in 1964; applying his methods to later years, the figure is seen to be 50 per cent in 1968.

It is reassuring that both value-added and asset figures agree in broad outline: a decrease from the early 1930's to the late 1940's; then a restoration; then an even further increase.

The trend can be embellished, for polemical purposes, by measuring from low to high, i.e., comparing 1948 with 1968, rather than from high to high (or at least earliest to latest), i.e., comparing 1931 with 1968. It is a bit like comparing department store sales, August with December, to show that sales are doing fine. Extrapolate the

1948-1968 "trend" by twenty or fifty years, and there is as much to view with alarm as there was decades ago; only a spoilsport would ask what happened to those earlier predictions of an imminent monopolistic economy.

The increased share of the largest 50 or 100, etc., has come about largely because the industries composed of big companies have grown faster than industries with smaller companies. (These industries of big companies, it should be noted, are *not* the industries with the highest concentration ratios. And within these big-company industries, there was actually some decrease in concentration ratios during 1947-1966.) The minor reason for the higher aggregate concentration has been mergers, some conglomerate some not, though such terms are highly imprecise.

But the most important question is the relevance of "superconcentration." Statistics do not speak for themselves. One needs a theory to show that a given set of numbers has a given meaning. A private-enterprise economy works well or badly through the network of markets which compose it. But there is no linkage between "aggregate concentration" and any market in the real world. At the limit, concentration ratios could be declining in every industry, and the less concentrated industries growing more rapidly, yet if those were big-company industries, "aggregate concentration" would increase.

The two biggest manufacturing companies are Standard Oil (New Jersey) and General Motors, with respectively \$17 billion and \$14 billion of assets. GM has about 55 per cent of U.S. automotive production. (The largest four automobile producers account for practically all domestic output. Yet as we warned earlier, a concentration ratio of 100 exaggerates. (GM, Ford, Chrysler, and American Motors do not in fact have the entire market. Imports are 13 per cent of car sales, and if we are to believe some auto executives quoted in the New York *Times* last May 22, imports will be 19 per cent if the industry does not change its offerings.) Jersey Standard, although bigger than GM, has only about 9 per cent of domestic oil refining, and less of production. (It is larger in the world market than in the United States, but the two are altogether separate.) *The huge size of a given company tells nothing about the rivalry it faces, the kind of market it lives in, or its price-output pattern.*

A useful mental experiment is to imagine that the largest 100 industrial corporations have all doubled in size and now account, not for about 50 but for 100 per cent of all manufacturing. In addition, they have completely diversified; every one of these companies operates in every market, in the same proportion as in total manufacturing. Then every market would have 100 sellers, of whom the largest would be only 7 per cent of the total. Few manufacturing markets show so little market concentration today.

Alternatively, suppose that 1,000 rather than 100 corporations account for all manufacturing activity, and that each one of the thousand is all alone in each of the 1,000 product classes (the so-called five-digit groups) enumerated by the Census. This would be, loosely speaking, universal monopoly. Strong competition with 100 firms and universal monopoly with 1,000 firms would seem to indicate that the number of the largest firms, regardless of industry, and the per cent of production which the largest firms account for, gives us no basis to say anything about competition or monopoly.

Assertions are made from time to time that somehow, somewhere there is a connection between bigness and concentration. A recent 730-page report by the staff of the Federal Trade Commission makes the assertion in many different ways; but one can find only one table which has some claim to relevance and at least asserts a slight connection between market concentration and the participation of the largest 200 firms. Were the table valid, it would mean that, e.g., an increase from 35 to 45 per cent in the participation of the top 200 in an industry would increase the concentration ratio in that industry from 42 to 44 per cent. The effect is too mild for discussion, and this is not the place to scrutinize technical defects, however grievous. But one general comment is unavoidable. The foundation of physical science is the reproducible experiment; that of economic statistics is the reproducible table. It is impossible to say how this particular table—like many other tables in the FTC report—was constructed. The lawyers would call it void for vagueness. And this is the sum total of evidence on the subject in the report.

The attention and publicity given to "aggregate concentration" derives partly from the belief that "the big firm" has basic advantages in the market place. It can outlast, outlose, outfight, outspend, etc., and thereby drive smaller rivals out of its respective markets without resort to anything so crude and costly as predatory warfare. (The belief is even stronger in Europe, despite the loss of export markets by American companies to much smaller European and Japanese firms.) The theory is general, and presumes to hold as well for the United States 70 years ago as for now. If it is correct, one should see an updrift in concentration ratios, as the firms which were largest in total size took over more and more of the markets in which they operated. The tendency would be stronger in some places, weaker in others, but over a long period of time it would be a tide lapping steadily upward and onward over the land. But as seen earlier, there is no such tendency; the chief problem is why industry concentration has remained so stable for so long, through stronger and weaker antitrust enforcement; through war, depression, and boom.

Plainly the theory is wrong, and size alone does not convey an inherent advantage. There are two mistakes in supposing that big and richer companies can elbow smaller rivals out of the market merely because they are big. First, investment for profit is not like spending for personal need or pleasure. If Mr. A and Mr. B both bid for the same painting, and Mr. A has ten times as much income and assets, he has the power to outbid and take the picture. But if Corporation X has a cash flow ten times that of Corporation Y, it has about ten times as many claims on it. There is no presumption that Corporation X can obtain more money more easily than Corporation Y for the particular purpose of improving its position in Y's market. Other markets may offer more lucrative opportunities—and such a comparison is bound to be made by management, the firm's creditors, and the investing public. Possibly the big firm can borrow more cheaply, though the advantage is never great and above a certain point, perhaps $50 million in assets, it goes to zero. Furthermore, most capital is equity capital, and the more profitable company can sell equity securities on better terms.

This brings us to the second and perhaps more basic mistake in supposing that larger companies can outlose or outbid smaller ones. Business firms only want money in order to make more. The firm which stands to gain more from an expenditure can offer more for the use of money. In any given market, therefore, the more efficient firm will tend to outbid its less efficient rival. The more profitable companies will tend to grow faster than the less profitable. If the process continues for some time, the biggest companies in any given industry may well end up being the ones who are the most profitable—but this hardly proves that size as such brings high profits.

But although "superconcentration" has nothing to do with market concentration, nor with monopoly and competition, it may be much more important. There may be strong *non*economic reasons for paying so much attention to the biggest companies simply because they are big. The feeling goes back a long way. A grandfather of the current Mr. Justice Harlan, who bore the same name, is best remembered today for his dissent in an early segregation case; his opinion that our Constitution is color blind was vindicated sixty years later. In 1911, shortly before his death, Justice Harlan wrote an opinion recalling the fear just after the Civil War that, with human slavery just abolished at appalling cost, there might be a new subjection to big corporations. There is a recent much-acclaimed biography of Huey P. Long, the brilliant and dangerous Louisiana Kingfish, who suited the action to the word that large corporations were wonderful to run against.

Public opinion has been, not hostile to big business, but at least cool and critical of it. It is no bad thing, in my opinion, that generations of big business executives have learned that they live in a goldfish bowl, more subject to taxes and regulations and prohibitions than smaller firms. But like all good things, this public opinion has its price, which is confusion and ambivalence, admiration and mistrust often equally misplaced. Like Huey Long, Mr. Dooley knew his America. The words he put into President Theodore Roosevelt's mouth would have been as true and apt twenty years earlier as they are now 65 years later: "Th' thrusts are heejous monsthers, built up be th'enlightened intherprize of the men who've done so much to advance progress in our happy country. On the wan hand, I wud stamp thim undher fut; on th'ither hand, not so fast."

Perhaps bigness is much more important, sociologically or politically, than is revealed by measuring economic quantities to understand market facts. If that is the case, it should be studied directly, and not be confused with economic concentration, a market phenomenon.

STRUCTURAL CHANGES IN THE ECONOMY

8

How obsolete is the business cycle?

R. A. GORDON

THE American economy has not experienced a serious business depression since the 1930's. It has shared this good fortune with the rest of the world. Indeed, in Western Europe and Japan even mild business recessions seem largely to have disappeared. Among the countries of Western Europe, North America, and Japan represented in the Organization for Economic Co-operation and Development (OECD), only the United States and, to a lesser extent, Canada continued until the 1960's to experience fairly regular business recessions of a mild sort: in 1948-49, 1953-54, 1957-58, 1960-61—and, history is likely to record in 1970.

The American postwar record of economic stability at high employment has been marred in another respect. For seven successive years, from 1958 through 1964, in business upswings as well as downswings, the annual average rate of unemployment did not fall below 5 per cent. After 1957, the monthly rate of unemployment did not decline to the long-sought figure of 4 per cent until the end of 1965, when the economy was beginning to feel the stimulating effects of accelerated military expenditures.

As the decade of the 1960's moved into its closing years, it seemed to many that the United States had entered into a New Era of rapid growth free of even minor recessions. Beginning in February 1961, a business expansion got under way which had still not clearly come to an end by the close of the decade. It experienced only a pause during the first half of 1967 which is now referred to as the "minirecession" of that year.[1]

If we do not identify a business recession in 1967, then by December 1969 the expansion that began in 1961 had lasted two months short of nine years.[2] This is far and away the longest cyclical expansion for which we have a documented record. The longest previous upswing (80 months) occurred during 1938-45, but this was dominated by wartime military expenditures. The longest previous peace-time expansion was only 50 months, from the bottom of the Great Depression in 1933 to the beginning of the brief but severe decline of 1937-38.

As the 1960's wore on, with the notable success of the tax reduction in 1964 and the growing prestige of the "New Economics," the question came increasingly to be asked in the United States, as it had been asked in Europe for some years previously: is the business cycle obsolete? Are even minor recessions a thing of the past?

Freedom from severe depressions

It had already been taken for granted, from the early 1950's on, that serious depressions were an extinct phe-

[1] When is a "minirecession" not a recession? For economists, the answer to this question is provided by the National Bureau of Economic Research. The Bureau, a highly respected nonprofit research organization, which is this year celebrating its fiftieth anniversary, has long been a leader in the field of business-cycle research. Part of its efforts in this area has taken the form of dating the turning points in American business cycles. These dates are taken as gospel—so much so that, for some years, no new dip in business has been regarded as a cyclical downswing by most observers until it has been officially confirmed by the National Bureau. In the first half of 1967, the Gross National Product (GNP), corrected for price changes, moved more or less horizontally for two calendar quarters; industrial production fell by about 2.5 per cent; and unemployment rose moderately to a little over 4 per cent. The Bureau declined to identify this pause in business as a "true" recession. Hence the title which has seeped into the economic literature on the 1960's: the "minirecession" of 1967.

[2] GNP adjusted for price changes stopped expanding in the fourth quarter of 1969, and a number of sensitive cyclical indicators began to decline in the third and fourth quarters of the year. At the time this was being written (June 1970) the question as to whether we had entered into a new business recession was still being debated. If the decision is eventually in the affirmative, the turning point may be put a little earlier than December 1969.

nomenon. Between the Civil War and World War II, the United States had suffered through seven "major" depressions. Four of them were prolonged: the declines of 1873-79 and 1882-85, the long depression of the 1890's, and the catastrophe of 1929-33. Three contractions were severe but relatively brief: those of 1907-08, 1920-21, and 1937-38. This confidence that severe depressions will not occur again rests on a number of grounds. These include widespread banking and financial reforms; the much greater importance of government spending relative to GNP than before World War II; the greater importance of the "automatic stabilizers"—for example, social security payments (particularly unemployment compensation), and cyclically sensitive corporate and personal income taxes—which tend to hold up disposable income and consumers' expenditures when GNP declines; the commitment to full employment at the end of World War II, a commitment that has grown stronger as continued prosperity has led to rising aspirations; and greater sophistication in the use of monetary and fiscal policy. This is not a complete list, but it is enough to lend convincing support to the belief that we need no longer fear severe depressions that drive the national unemployment rate up to 10 per cent or more. (The figure was 25 per cent in 1933.)

Severe depressions do, indeed, seem to be a thing of the past. But can the same be said of the minor recessions such as the United States experienced between 1945 and 1961, and which had recurred with depressing frequency as far back as the National Bureau's records go? Did something happen in the 1960's to make the business cycle obsolete in its milder as well as its more severe form? This subject is being debated anew as this is being written.

Redefining the business cycle

Against this background of unprecedented freedom from cyclical instability, some forty economists from Japan, North America, and Europe converged on London in 1967 to debate the question: "Is the business cycle obsolete?"[3] The consensus among these experts was: No,

[3] The proceedings of this conference have recently been published. See Martin Bronfenbrenner, ed., *Is the Business Cycle Obsolete?* (New York: John Wiley and Sons, 1969). A significant volume that deals with substantially the same range of questions is Erik Lundberg, *Instability and Economic Growth* (New Haven and London: Yale University Press, 1968).

it isn't, but we need a broader definition of the cycle. Particularly so far as Europe and Japan are concerned, and this may be becoming increasingly true of the United States also, we have to think of business cycles as not only cumulative ups and downs in the *absolute level* of total output and other important variables but also as cycles of acceleration and retardation in the *rates of growth* of these variables. Periods of retarded growth can result in rising unemployment, declining profits, and other characteristics that we associate with a business recession. But, the rise in unemployment and other recessionary features of the retardation phase of a "growth cycle" would be milder than in a recession in which total output shows an absolute decline.

Such retardations in growth, or "growth recessions," occurred in much of Western Europe several times in the postwar period—in 1951-52, most clearly in 1957-58, and to a lesser extent in 1961-62. In addition, some countries had "growth recessions" not shared by others. A succession of "stop-go" policies stemming from balance-of-payments difficulties has brought the growth of the British economy to a virtual halt on several occasions. Even fast-growing Japan experienced a number of these "growth recessions" in the postwar period, as Italy did in 1964, and Germany in 1966-67.

Government policy has played an important role in a number of these cyclical setbacks. Restrictive policies have been used not only in Britain but also in other countries to control inflationary booms, sometimes (but not always) because of threats to the balance of payments. This has led some economists to speak of "policy cycles." Restrictive policies have helped to bring on brief periods of retarded growth and rising unemployment—and sometimes belated reversals of policy have exacerbated the booms that followed. To some extent, therefore, governments have contributed to a mild form of cyclical instability. But this has been more than offset in most advanced countries by the contributions that governments have made to both short-term stability and long-run economic growth. Official commitments to the goal of full employment, the increased ratio of government expenditures and transfer payments to GNP, the increased willingness to use—and greater sophistication in using—the instruments of monetary and fiscal policy have all tended to reduce the amplitude of cyclical fluctuations and, at least in Europe and Japan, to convert the old-fashioned business cycle into the "growth cycles" of the 1950's and 1960's.

Dynamic properties of the American economy

To suggest that the United States will experience little or no significant cyclical instability in the future is to imply that notable changes have been occurring in the dynamic properties of the American economy. Where have such changes been taking place, particularly in the last decade?

Today the economist's way of dealing with questions of this sort is to construct a model of the economy, as exemplified by the numerous econometric models that have been developed for this and other countries. These models seek to express mathematically the relationships that are presumed to govern the dynamic behavior of the economy. An econometric model, in effect, considers the behavior of the economy to be determined by three sets of influences:

(1) How the behavior of the *endogenous* variables (i.e., variables whose behavior the model seeks to account for) is explained by specified relationships—for example, how consumers' spending relates to the income and the wealth of households or how wage changes relate to price changes, unemployment, and other factors.

(2) *Exogenous* variables which the model does not seek to explain, such as government spending, and

(3) Random shocks or disturbances reflecting the multitude of influences that the model does not succeed in identifying.

In terms of such a model we can restate the question as to whether the old-fashioned sort of business cycle is extinct, or on its way to extinction, in the United States:

(1) Have there been in the last decade or so important changes in the nature of the relationships determining the behavior of the endogenous variables—e.g., consumption, business investment, the levels of prices and wages, and so on?

(2) What changes have occurred in the behavior and relative importance of the exogenous variables? For example, is government spending more stable in the short-run than in the 1950's, and is the ratio of government spending to GNP significantly larger than it was?

(3) Have there been any significant changes in the

pattern of random shocks to which the economy is continuously subjected?

Changes in economic relationships

Recent research does not suggest that there have been many significant changes in the nature of the dynamic relationships that govern the behavior of the important endogenous variables. Indeed, current econometric models are based on statistical calculations for the period back to the Korean War. These assume the relevant relationships remained unchanged in the 1960's.

To cite a few examples. There is little evidence that the various types of consumers' expenditures respond to changes in current disposable income in any fundamentally different way from that in the 1950's or earlier. True, most economists were surprised at the failure of consumer spending to respond more sensitively than it did to the tax surcharge in 1968, but this is generally taken to reflect our imperfect knowledge rather than any significant change in underlying relationships. This experience in 1968 was, by the way, something of a blow to the "New Economics." Among other things, it suggested that moderate changes in tax rates which are assumed to be temporary have relatively little effect on consumer spending, in contrast to the successful tax cut in 1964 which was announced as a permanent reduction.

At the heart of all business cycle theories is the volatile behavior of business investment in plant, equipment and inventories. Minor business recessions in the United States have been particularly associated with swings in inventory investment. The experience of the 1960's does not suggest that the destabilizing role of inventory investment has disappeared. It is true that through 1961 there had been a tendency for the cyclical swings in inventory investment to diminish in relative importance. But in 1966 there was an upward surge in inventory investment, to be followed by a precipitous decline in the first half of 1967. Had it not been for the rapid increase in military expenditures and the relative stability of private fixed investment, we should have recorded a true recession in 1967 rather than a "minirecession." Despite improvements in inventory management and a downward trend in the inventory-sales ratio, wide swings in the rate at which inventories are accumulated can still occur.

Empirical research does not suggest that the factors determining the short-run behavior of business expenditures on plant and equipment have experienced any significant structural change in the last 10 to 15 years. Even during the 1950's it was assumed that large-scale business was using a longer time-horizon in its investment planning and that consequently fixed investment was becoming somewhat more immune to changes in current business conditions. This continues to be true. But this tendency does not guarantee that business investment will not fall in response to declines in output or profits or other influences making for cyclical contraction.

It is true that surveys made at the end of 1969 and in early 1970 indicated that business firms were planning a surprisingly large amount of plant and equipment expenditures for 1970, despite the accumulation of unfavorable news. A partial explanation is probably the widespread assumption of a continued rapid rise in prices. While helping to support the level of economic activity in 1970, the continued investment boom may be storing up trouble for the future. (After this was written it began to be evident that business was revising downward its planned investment for the rest of the year.)

The behavior of residential building has served as an important countercyclical force in the postwar years. Tightening credit conditions during a business expansion cause residential construction to decline. The easing of credit that comes with a business downturn leads to a revival of building. In 1969, as in 1966, a shortage of mortgage funds led to a sharp contraction in housebuilding. An easing of credit following on a subsiding of inflationary pressures will lead to a revival of residential construction and thus tend partially to counteract recessionary tendencies in other parts of the economy. But this was true also in the 1950's and earlier, when minor business recessions were a regular occurrence.

Wages, prices, and inflation

There is one respect in which the cyclical behavior of the economy does seem to have been changing, but this change became evident as early as 1953-54. The business contraction of 1948-49 was our last recession in which there was a significant decline in wholesale or retail prices. In the recessions of 1953-54, 1957-58, and 1960-61 the wholesale and consumer price indices remained substantially level or actually increased. In short, prices have become more inflexible in a downward direction.

This observed change in the cyclical behavior of prices raises another question that many are asking now. Is the economy today more susceptible to inflation than in the past? The question requires more than a simple "no" or "yes" answer. During 1961-65, before the upsurge in military spending for Vietnam, economic expansion was accompanied by a modest rate of increase in the Consumer Price Index; wholesale prices moved horizontally through 1964; and wages rose less than past experience had led economists to expect. One result was a rapid rise in profit margins and total profits. This interrelated behavior of wages, prices, and profits *was* unusual for a business expansion lasting as long as four to five years. Typically, costs have begun to encroach on profits after a business expansion has been underway for two or three years.

Acceleration of the rise in prices began in 1965. Wage increases accelerated with the decline in unemployment to below 4 per cent in 1966 and were further stimulated by the rapid rise in consumer prices during 1966-69. At the same time, profit margins began to sag.

There is nothing new in this sort of interrelated behavior of wages, costs, and prices. Initially the inflation of 1965-69 was of the "demand-pull" variety, fed by the rise in government spending and by the largely induced expansion in both consumer expenditures and business investment. Rising prices and low unemployment in turn generated accelerated wage increases, which then became the basis for still further price increases. (The "cost-push" effect.) There is nothing new about this sort of interaction.

What seems to be new, at least for the United States during a period free of wartime controls, is the length of time during which this interaction went on. Unemployment remained below 4 per cent for four years. This is the longest period in this century in which unemployment has remained this low without being associated with wartime controls. Annual average unemployment remained below 4 per cent during 1943-48, but wartime rationing and price and wage controls existed during 1943-45; and the next three years saw a rapid upsurge of prices as wartime controls were relaxed and pent-up demands were released.

The unusually long period during 1966-69 of low unemployment and excess demand has, to put it loosely, generated a momentum which is difficult to stop. Even though unemployment began rising early in 1970, wages were still reacting to the accelerated price increases and low unemployment of the past years. And these wage increases were to a substantial extent being passed on as price increases.

The classical prewar method of dealing with such a situation was a strongly deflationary policy aimed at reducing the level of aggregate demand and eliminating expectations of further price increases. Here we do face a changed situation from that in the 1950's. It is unlikely that the administration dares risk the unemployment levels reached in previous postwar recessions. Also we may be experiencing some shift in the way wage changes are related to the level of unemployment, a relationship economists refer to as the "Phillips curve." It seems too early to say. Recent experience does suggest, however, that wage inflation is related not only to the current level of unemployment and to current and recent price increases but also to the length of time the economy operates at a full-employment level or higher. We do not have the experience yet to determine what rate of inflation is likely to be associated with *prolonged* periods of full, or better than full, employment in the United States.

One thing it is safe to say. Prolonged maintenance of an unemployment rate below 4 per cent will involve a rate of increase in the price level significantly higher than that which we experienced during 1958-64, when the unemployment rate remained above 5 per cent—and also higher than during the boom of 1955-57, when unemployment hovered around 4 per cent for a little over two years.

Monetary policy and the money supply

As noted earlier, we need to look also at the behavior of the more important variables that we take to be exogenous. Here, two key policy variables may be identified: the supply of money and government spending.

During the first half of the 1960's, the Federal Reserve authorities permitted the money supply to increase at a moderately accelerating rate as they supported the efforts of the Kennedy and Johnson administrations to move toward full employment through the use of an expansionary fiscal policy. It was certainly true in the mid-1960's, as it is now, that the Federal Reserve Board took the goal of full employment more seriously that it had in 1959-60, when it began to contract the money supply while the unemployment rate was still above 5 per cent.

Since 1966, however, there have been wide swings in the rate of increase in the money supply. The Federal Reserve blundered badly in 1968, when it permitted the supply of money to increase rapidly on the mistaken assumption, shared by many others, that the tax surcharge would apply an effective brake to the economy. Then, in the second half of 1969, the expansion of the money supply was brought to a complete halt, and interest rates soared to heights not reached before in this century.

Milton Friedman and his followers have had a strong influence on economic thinking in the last few years. Today virtually everyone is prepared to concede that "money does matter." But the channels through which changes in the supply of money affect the economy are still being debated; the lags involved are long and variable; the tight money policy of 1969 seems to be having a distressingly slow and so far slight effect on the price level; and we are not yet sure what the full effect will be on the levels of output and unemployment. To all this we must add the inevitable errors in analysis and prediction which will continue to afflict the monetary authorities, who after all are merely human.

The Federal Reserve is less likely now to permit monetary tightness to last too long than was the case on some past occasions, but already in the spring of 1970 it was being criticized for not easing credit conditions faster than it did. The Federal Reserve Board, as well as the Nixon Administration and the Congress, must make its choice among various combinations of degrees of inflation and unemployment. Its choices—and its imperfect knowledge—can lead us into recessions in the future. Indeed, it can be argued that this has just occurred.

Government spending

Total government expenditures on goods and services (federal plus state and local) were a significantly larger percentage of GNP in 1969 (23 per cent) than in the mid-1950's (about 19 per cent). Interestingly, because of the much larger GNP, federal defense spending was actually a smaller fraction of GNP in 1969 than in 1956. Virtually all of the increase in total government share of GNP between 1956 and 1969 was made by state and local governments; the federal share, including both defense and nondefense spending, remained essentially unchanged. The rapid growth in state and local government expenditures has provided important support for the growth in aggregate demand. Further, state and local spending has shown less erratic changes than have federal expenditures.

It is apparent then, that federal spending on goods and services has *not* been a stabilizing influence in the postwar period. As Bert G. Hickman complained, in reviewing postwar experience through the 1950's, federal spending is the least stable of the major components of GNP. The same can still be said. This instability can be traced primarily to the dominant role played by defense expenditure. Federal spending will continue to be a major source of instability so long as it consists predominantly of military outlays.

The deflationary consequences of a future large reduction in military spending could be offset to a considerable extent by a vigorous use of discretionary fiscal policy—through quickly enacted tax reductions and a prompt and substantial increase in nonmilitary expenditures. It is still true that the present level and composition of federal expenditures and their probable future behavior permit, and even make probable, some degree of cyclical instability in the economy.

Stabilization and the "New Economics"

In reviewing the difference between the behavior of the economy in the 1960's and in earlier postwar years, Arthur Okun, the last chairman of the Council of Economic Advisers under President Johnson, recently wrote: "The nature of economic fluctuations has not changed; policies to contain them have made the difference."[4]

The policies that helped to keep the 1960's free of recessions involved more than the discretionary use of fiscal measures. In the words of Arthur Burns, the central notion was "that the stage of the business cycle has little relevance to sound economic policy; . . . that the vital matter is whether a gap exists between actual and potential output; that fiscal deficits and monetary tools need to

---

[4] Arthur M. Okun, *The Political Economy of Prosperity* (Washington, D.C.: The Brookings Institution, 1970), p. 35.

be used to promote expansion when a gap exists; and that the stimuli should be sufficient to close the gap—provided significant inflationary pressures are not whipped up in the process."[5]

In other words, expansionary measures are called for if real GNP is below the level that would yield full employment (that is, potential output), even if a business expansion is already under way. This program was intended to be symmetrical—which is to say, deflationary measures are called for if aggregate demand drives GNP above potential output, with a consequent accelerated rise in prices.

Another foundation stone of stabilization policy in the 1960's was the concept of the "full-employment surplus"—that is, the budgetary surplus or deficit that would exist at full employment. Thus the tax cut of 1964 was made in the face of a *current* budgetary deficit because, it was argued, the prevailing level of tax rates and the sensitivity of tax revenues to changes in GNP were such that, without government intervention, the expansion then under way could not reach full employment because of the restraining effect of rapidly rising tax revenues.

Discretionary fiscal policy, when faced with the volatility of military expenditures and the brute realities of the political process, could not prevent wide swings in the "full-employment budget" in the latter half of the 1960's. Fiscal (and monetary) policy failed to prevent the build-up of inflationary pressures during 1966-68, and fiscal and monetary policy together then brought on the squeeze which stopped economic growth and led to rapidly rising unemployment in the early months of 1970.

This is not to suggest that important gains have not been made in the area of stabilization policy. Emphasis on the "full-employment budget" and on the concept of "potential output" is continuing under the Nixon administration. Without doubt the Federal Reserve authorities have learned a good deal from the mistakes of the last few years, including the need to take more specific account of the rate of change in the money supply. And emphasis on the goal of full employment is continuing to increase. An unemployment rate of 5 or 6 per cent seems much less tolerable today than was the case in the Eisenhower years or earlier. Indeed, an interesting fact should be reported that has received little notice. The first *Economic Report*

[5] Quoted approvingly by Okun, *ibid*, p. 43.

of the Nixon administration, published in February, 1970, defines potential output as that which the economy is capable of producing "at an unemployment rate of about 3.8 per cent"—thus implicitly raising the full-employment goal from the "interim target" of 4 per cent used during the two previous administrations.

New labels for old

To quote Arthur Okun again, the "experience of the sixties has made a marked and lasting change in business cycle mentality." He cites the fact that the widely-used government publication presenting current economic indicators changed its title from *Business Cycle Developments* to *Business Conditions Digest*. It is also true, as he mentions, that "Business Cycles" is no longer the title applied to most of the current research and teaching concerned with the dynamic workings of the economy. "Business Cycles" is tending to disappear from textbooks and courses, and teachers and students now concern themselves with a field that bears such titles as "Economic Dynamics," "Growth and Stability"—or "Growth and Instability."

Be that as it may. We continue to live in an economy the behavior of which is governed by a complex set of dynamic relationships which generate both long-term growth and short-term cumulative movements upward and downward. Sufficient structural change has occurred so that we can be reasonably confident that these cumulative movements will not push us into a serious depression. With the postwar emphasis on full employment and rapid growth, the increased role of government, and growing sophistication in the use of stabilization measures, we can also hope to decrease the frequency and amplitude of minor business recessions.

One price we shall have to pay, and are paying, is a degree of persistent inflation. And efforts to curb such inflation can lead us into new—but hopefully brief and mild—business contractions.

Before World War II, the Keynesian Revolution, and the construction of large-scale economic models, economists studied a phenomenon called "the business cycle" with only very general notions as to the nature of the complex of dynamic relationships that generated cyclical fluctuations. Now the emphasis is on trying to specify the nature of these dynamic relationships, what they imply as to how the economy is likely to react to various sorts of disturbances, and how policy can help to control and offset cumulative responses of a sort that we should like to avoid.

We still have much to learn. Even if we knew more, we could not eliminate the shocks to which the economy is prone; nor are we likely soon to overcome the political and administrative difficulties in applying the knowledge that we do have. So it is not likely that we have seen our last recession—or our last inflationary boom. Indeed, we moved into a "growth recession" toward the end of 1969, which can also be said to mark the contraction phase of a "policy cycle." It remains to be seen whether the National Bureau of Economic Research will eventually report that, according to its criteria, we also have been witnessing the first "true" business recession since 1961. I think that we have.

9

Whatever happened to British planning?

NORMAN MACRAE

THIS is the story of the honourable birth, unsuccessful life, and present coma (it may not yet be death) of an originally admirable political spoof word. I say "admirable," because the popularisation of the possibly meaningless word "planning" in Britain's political vocabulary in the dear, dead 1960's did not start as the result of an attempt by politicians to fool the people. It started as the result of an attempt by economic writers to fool politicians in a good cause.

As I was at the time writing many of the leading articles on enomic affairs in *The Economist,* I played a part in the conspiracy myself; and now seems to be the time to come clean. Obviously, those who were on the other side of the hill during the verbal struggle will not agree with a word I say here. Maybe half-rightly; perhaps they were successfully half-fooling us at a time when we thought we were wholly fooling them. But here is how things looked from the guerrillas' encampment.

The story begins at the opening of the 1960's when Britain's Chancellor of the Exchequer was a nice, cautious unimaginative Conservative called Selwyn Lloyd. It was once cruelly said of Selwyn Lloyd that his main hobbies were hard work, gardening, and doing what he was told.

At this time he was being told by the civil servants at the Treasury to deflate demand within the British economy, especially at any period when Britain's balance of payments got into difficulty. It was rather odd that this was such united Treasury advice, because in only two of the previous eight years had internal demand in the British economy in fact increased by more than productive potential: these two years had been the election years of 1955 and 1959 when Conservative Chancellors had given away large tax reliefs because they wanted to win votes. But it had become an article of faith with the civil servants that those two wicked pre-election booms had caused all Britain's troubles. They were determined to sit hard on the Chancellor's head, and to tell him to keep down demand.

While it is fair comment that the Treasury civil service was too much in the hands of people who held this view, it is no doubt equally fair comment that the non-mass media—by which I mean the opinion-forming press and those who had access to current affairs programmes on the television screens—were too exclusively occupied by people who hold the opposite or "expansionist" view. I was unashamedly of this company.

We "expansionists" believed that the main reason why Britain's balance of payments was likely to go increasingly wrong in the 1960's was that the sterling exchange rate was too high. After all, when the pattern of world exchange rates had last been fixed in 1949, at least four of the main industrial powers of the world (Japan, Germany, France, and Italy) were still largely war-destroyed. The two big industrial powers who were not war-destroyed, America and Britain, were therefore bound to find within the next twelve years or so that their exchange rates were too high, relative at least to the currencies of Japan and Germany. Even if Britain had followed a brilliant economic policy between 1949 and 1961 (which was far from being the case), the relative strengthening of the economies of Japan and Germany in that period was almost bound to be greater than the relative strengthening of Britain's economy; in 1949 Japan and Germany had nowhere to go but up. This was one reason why some of us had urged Cripps back in 1949, and then Butler in 1952, to float the sterling exchange rate: because we believed that by the end of the 1950's sterling (and the dollar) would be overvalued relative to its main competitor currencies, the mark and yen.

Because of the overvaluation of their currencies, it

seemed probable that both America and Britain would run balance of payments deficits in the 1960's. To America, it was clear, this need not matter a damn. The United States could pump out dollars to pay for its deficit, and foreign countries (sometimes complainingly) would be willing to hold them. But if Britain pumped out sterling to pay for its deficits, thus adding to the large sterling balances that were already unwillingly held, people would simply run from sterling. As these runs from sterling mounted, the danger was that Treasury civil servants would tell British Chancellors to deflate the British economy a bit more. Not gravely deflate it; none of us feared a serious recession with high unemployment. But we did fear that British policy would be kept constantly stern enough to stop any healthy growth in production, and particularly any growth in investment. Unlike the Treasury, we did not think that this bit of deflation would promptly drive British manufacturers into exporting more; deflation can divert supplies from the home market to exports when exports are profitable, but not when an overvalued currency means that marginal exports even in squeeze-time are sent out at a big loss. And if deflationary policies lasted in Britain throughout most of the 1960's, the delay to new investment would mean that the country's already outdated stock of capital equipment would be even more outdated by the end of the decade. I would add that this is not just what we feared would happen to Britain during the 1960's. I would still argue that it is precisely what did happen to my country during most of that rotten period.

So, at the beginning of the 1960's, the theme songs of expansionist economists were to urge Selwyn Lloyd (a) to go for growth; and (b) if he ran into balance of payments troubles to devalue the pound. It is a fair criticism that in the press we called for (a) more determinedly than (b). The reason was that whenever any of us mentioned the case for devaluation—and this was inevitably most newsworthy at times of trouble—we were made to feel like Benedict Arnolds. The Governor of the Bank of England or some other dignitary would call us, or our editors, or our newspapers' other bosses, and say (usually quite falsely) that the drain on Britain's gold reserves the previous week had been largely due to something frightfully irresponsible that one or other of us had written about the pound.

Meanwhile, the Treasury regarded as irresponsible even the calls of "go for growth" which the press was sending up. Interestingly, however, several hints began to percolate through that Prime Minister Macmillan did not; he, too, was wondering whether a policy of constantly restricting economic growth in order to try to save the exchange value of the pound was wise. It therefore seemed that it would be a very good thing if some prestigious body—far more prestigious than the press—could come into being in order to shout into the Chancellor's left ear "go for growth," even while the Treasury continued to murmur into his right ear "disinflate demand." And it was at this time that articles began to appear in the British press saying that France had owed a great deal of its economic advance in the 1950's to the brilliance of French economic planning, and to the existence in France of its independent *commissariat du plan.*

I have no doubt that some of those who wrote these articles really believed in the efficiency of French planning, but others of us frankly used discussion of it as a heaven-sent weapon in the battle to get the Treasury unseated from the position of sole and supreme power. The British Government was at this time eager to flatter France, because it was beginning (rightly, and here the officials at the Treasury were wholly on the side of the angels) to start to want to get into the European Economic Community. It followed that when it was widely suggested that Britain should study the possibility of setting up its own version of the *commissariat du plan* even the most crusted of Treasury mandarins could not tactfully give the usual British civil service reply of "we're not going to learn from those damned frogs." Moreover, the founder and first head of the French *commissariat du plan* had been Jean Monnet, the revered father of the common market who was greatly admired by all the best people in Britain: including both the expansionists and the Treasury mandarins. Finally, Macmillan himself was rather enamoured with the words "economic planning"; he had written in favour of them in his youth, without being quite sure what they meant. So the idea was sold that Britain should at least study whether it should set up its own version of the French *commissariat du plan:* no doubt a watered-down version, but an imitation in some shape. Towards the end of 1961, parties of civil servants, politicians, and journalists went over to Paris in a steady stream to study what the prototype was. What did we find?

The simplest way of describing the planning mechanism

in France is to say that it has had two objects. The major one was necessary mainly because France had not got a proper capital market. Frenchmen's habit of putting their savings into gold and peasants' stockings and other idle resources helped to make it respectable for the French government to create funds for industry equal to those idle savings, and lend them out through various official and quasi-official agencies. In order to find out who best to lend them to—all right, this is a simplification but not much—the French authorities arranged for various studies of industrial projects, and held various conferences between appropriate groups of industrialists, trade unions, and others, with the *commissariat du plan* doing some of the job of project studying. It is true that the French system also allows for discriminatory tax treatment of firms that are doing what the public authorities want, but it is this planning of the best use of scarce supplies of credit that has been the planning commission's most important job. In effect, it has worked out how funds would have been distributed by a developed free capital market of the American or British type if a developed free capital market existed in France.

Since Britain has, in the City of London, a free capital market of the most experienced type, there was no need in London for a planning commission with these functions. Anyway, the City would not have allowed one to work in this way if any government tried to set one up. Oddly, however, it transpired that Selwyn Lloyd liked the idea of meeting in a rather pointless new council with industrialists and trade unions, and setting up bodies to discuss the affairs of different industries. Since such bodies would not have anything major to do, the Treasury had no objection.

But it was the second function of the French planning authorities which the Treasury's opponents were conspiring to establish in Britain. The *commissariat du plan* regularly gave publicity to an economic growth rate which France ought to be able to attain; and the French Treasury and the Bank of France drew up their policies in the knowledge that they would be regarded as failing in their job if something like that growth rate were not attained. This idea of a target growth rate, equal to the growth of productive potential, was commonplace all over the world by 1961-62. In America, Walter Heller was to say that the main achievement of his own Council of Economic Advisers was that it propagated the

> concepts of economic potential and the G.N.P. gap and the targets of 4 per cent unemployment and 4½ per cent annual growth. All of these were imprinted with the Presidential seal in Kennedy's first year and thereby became administration policy.

As a more extreme example, in Japan there are actual debates within the cabinet whether the published target growth rate for the year ahead should be, say, 10.2 per cent or 10.3 per cent. This has a real meaning in Japan because of their strange budgetary system which aims at exact balance of the budget at the rate of national income (and thus of tax revenue) that is assumed in the forward projection. If the government sets its target growth rate 0.1 per cent higher, this means that there is that much extra revenue assumed to be coming into the kitty—and available for giving away in tax reliefs or for spending on some desirable form of government expenditure. Even in Germany at this time (allegedly the land of no planning), the visiting journalist was readily told the assumed rate of economic growth for the year ahead, on which the revenue estimates in the budget were based. Only in Britain, which was supposed to have started Keynesian economics, were target growth rates for the year ahead kept a deadly secret in the early 1960's. There had to be such assumed growth rates behind the scenes, because the published estimates for government revenue were based upon them. But the Treasury civil servants told Chancellors to keep them very quiet. Otherwise, it was feared, the wretched journalists would stir up political trouble by saying that the target growth rates were too low; and this would divert public attention from what the Treasury thought was the main objective of "keeping sterling strong."

IN THE first half of 1962, Selwyn Lloyd gave his approval to the establishment of a National Economic Development Council, promptly nicknamed "Neddy." The council itself was to be a talking shop between ministers and respectable industrialists (including leaders of nationalised industries), trade union bosses, and establishment-type public figures. It was to sprout some study groups of particular industries, fairly promptly nicknamed "Little Neddies," but without real powers. However, Neddy was also to have a backroom staff, whose first head was Sir Robert Shone, a respected business economist who believed in economic growth. It was on Robert Shone that we expansionists pinned our hopes. We were confident that his reports would say that a target growth rate of 4 per cent a year in Britain should be given some sort of priority of its own. This did in fact happen, even although it was a bit disappointing that the staff of economists he hired for Neddy were mostly microeconomists (i.e., students of the market) rather than macroeconomists (i.e., students of how much demand to allow into an economy).

However, before Neddy even took its first steps, a dramatic political development made its most important function temporarily unnecessary. In mid-1962, Prime Minister Macmillan sacked Selwyn Lloyd from his post as Chancellor of the Exchequer. Macmillan had his eye on the general election that was due in 1963-64. He wanted tax reliefs and economic growth in 1963 to help win that election, and rightly suspected that Selwyn Lloyd would obey Treasury advice to keep on a deflationary tack instead. Reginald Maudling, the new Chancellor, was an expansionist; but he obeyed Treasury advice in some matters too. When Robert Shone produced his first full report from Neddy, there was a passage in it discussing the case for devaluation of the pound. Maudling insisted on its being struck out before the report was published. By this time, however, most of us were not interested in the activities of this dying Conservative Government. The question was: what would an incoming Labour government do about planning? For the public opinion polls made it fairly clear that before the end of 1964 a Labour government would be in power.

Harold Wilson's attitude towards economic planning was absurdly influenced by the fact that George Brown and James Callaghan had been his two opponents in the vote for the succession to Gaitskell in early 1963, and he felt he owed a chivalrous victor's duty to both. Jim Callaghan had been promised that he would become Labour's Chancellor of the Exchequer, and was busying himself reading books about tax reform. George Brown wanted the Foreign Office, but it was felt that George would not be very good at stuffy diplomatic cocktail parties. Brown's next choice was "something in the economic-industrial field"; and this was why Harold Wilson decided to set up the Department of Economic Affairs, just down the Whitehall corridor from the Treasury, with the idea that two departments running British economic affairs would cause some "creative tension."

Wilson later claimed that the thought of another department as counterweight to the Treasury had been recommended years before by outside economic writers; and was cross with those of us who slanged our own dreamchild. In the days when we were desperate to get another voice speaking up for expansion, and before the invention of the ploy *à la Parisienne* called Neddy, no doubt some of us had written in favour of something of the sort. But, in the context of a returning Labour government, against which the conservative hackles of the Treasury civil service were already aroused, the idea was a crazy one. From the moment it became known that Wilson would create a DEA, the Treasury closed its ranks to make quite sure that this upstart down the corridor did not steal any of its duties and privileges.

The Treasury had been asked to provide some staff for the new DEA. It sent over mainly its "long-term" thinking group, which had no connection with immediate policy and which some people had long thought of disbanding, although its exports of staff to the DEA were also carefully diluted with men who would be sure to owe their main duties of loyalty back to the Treasury with whom (under the peculiar British civil service system then current) their hopes of future promotion anyway lay. Apart from these Treasury *emigrés* and a few outsiders, the new DEA was largely recruited by bringing in the microeconomists from Neddy. This left poor Robert Shone at Neddy in charge of nothing except a federation of little Neddies, engaged in unimportant studies of particular industry; and even over these his job was diminished, because George Brown had brought an industrialist friend called Fred Catherwood into the DEA, with the job of changing some of the ways in which the little Neddies might work. Catherwood later succeeded Shone at Neddy, which is now a talking shop where the Prime Minister (as well as the economic ministers) meets establishment figures from the industrial and trade union world at odd intervals, and in which little Neddies tend to be pressure groups seeking favours for their industries.

But the Treasury's main concern, as Labour came to office in October 1964, was to rally round its new Chancellor Jim Callaghan, and show that he had the power over Britain's economic policy, not Brown and the DEA. It was partly for this reason—ridiculous though it may sound—that Callaghan introduced a full-scale emergency budget debate in Labour's second month of office.

In this Callaghan announced that he was going to raise old age pensions, and pay for this by a rise in income tax; he was also going to introduce a new corporation tax and a capital gains tax. This set of measures, besides indicating that Callaghan was the economic boss, sounded so "socialist" that there was naturally a large run from the pound. The Bank of England then borrowed the money to prop up the pound; and the defenders of proper channels in London (but nobody else) breathed easily because the old team of the Treasury and the Bank were now clearly again the masters in Britain's mortgaged house.

At the DEA, George Brown was left to busy himself with two things. One was incomes policy, by which he was supposed to try to keep down the rises in wages granted to the trade unions. Because he was given no powers with which to enforce such a policy, it was bound to fail. Brown persuaded the trade unions and employers to sign a "declaration of intent" that they would try to exercise moderation, but the unions had no intention of honouring their word. Secondly, Brown plunged into the preparation of his so-called "national plan." This was a pretty contemptible farce.

The Treasury had agreed that a target rate of just under 4 per cent per annum economic growth could be written into this paper plan. The DEA then wrote to people who were supposed to be representative of various industries, and asked them what they thought their own growth rates would be if something near to 4 per cent were the national average. Naturally, industries that were in fact due to decline said that they thought their growth would be just under 3½ per cent per annum; nobody wished to say that his industry would waste away, because it was felt that if one had a positive growth rate written into the plan perhaps one would be able to get some government help when economic forces began to oblige one's factories to close. At the other extreme, industries that were obviously liable to expand by over 10 per cent a year said that perhaps they would expand at just over 4½ per cent; nobody wished to look as if he expected to be too prosperous, in case Callaghan came in with a special discriminatory tax.

The microeconomists who had come to the DEA from Neddy then pored over these replies from the industries, which had been compiled with differing degrees of seriousness and care (or lack of it). The main worry of the microeconomists seemed to be that the figures should add up arithmetically. Once they did, the document was published as the so-called "national plan" in September 1965. It was 270 pages of rubbish in every respect except one. The saving grace was that the Government had taken some care in filling in the parts of the questionnaires that related to its own intended spending. Probably the exercise over the national plan helped to inculcate into Whitehall new methods of forecasting and controlling government expenditure: a small achievement, but worthwhile.

However, the "national plan" was much talked about during Labour's successful election campaign in 1966; and George Brown genuinely believed that he had sponsored something very important. His argument was that a 3.8 per cent annual growth rate was now written into the plan as stated government policy. If another international economic crisis blew up, Labour was now committed not to counter it by departing from 3.8 per cent growth. The rest of us thought this view was too optimistic. Sadly, we proved to be right, and Brown to be crashingly wrong.

Four months after Labour won the general election of March 1966, there was another run from sterling. It came about largely because of America's "credit crunch" in the summer of 1966. American banks were pulling back to New York every dollar they could, and this induced other people to switch out of sterling into dollars on a large scale. George Brown told the cabinet that they must meet the crisis by devaluing the pound. The cabinet refused to do so. On Treasury insistence Jim Callaghan announced another heavy dose of attempted disinflation in order to "make sterling and Britain strong." Brown decided to resign, and then withdrew his resignation. But he moved away from the DEA as soon as he decently could. Wilson shifted into DEA two successive ministers who were sure not to be active enough to annoy the Treasury. Last year the DEA was finally abolished altogether.

Nobody talks of the national plan any more. There was, prior to the recent Conservative victory, some talk about "restructuring" industry, because Labour had given new funds to the Ministry of Technology and to the so-called Industrial Re-organisation Corporation (which is supposed to sponsor mergers). Naturally, there are claimants for these funds. But MinTech had used most of its funds to prop up concerns that it would have been more economic to allow to die; and the IRC had sponsored some dubious mergers as well as some that would have occurred in the free market anyway. On balance, government "interventionism" since 1964 has served to push resources slightly in the wrong directions, not in the right ones. The new Heath government will be quite right to abandon this sort of "planning" while maintaining the one advance that Labour "planning" has left behind it.

It is in the Treasury itself that there has been an advance. When sterling was at last devalued in November 1967—at least seven years late—Roy Jenkins came in as Chancellor of the Exchequer. Jenkins had been an advocate of growth and devaluation long before: a member of our company of expansionists in his free lance journalist days. Once in the Treasury, he did not "go for growth" as ardently as the wilder expansionists had hoped. The fact that he was a bit tamed by the Treasury probably shows that many other economists would be tamed if in the same position. However, Jenkins did institute one really important reform. At the same time as he presented his annual budget he published each year a separate paper showing the economic growth rate at which his policies aimed for the year ahead. Even if the target figure was disappointingly low, it was now a stated target; this meant that everybody working within the Treasury nowadays has some quantitative grip on at least what he thinks he is trying to do. This is a real change in practice; economic policy is now directed with an aim in view, instead of just being a matter of periodical kicks at the brake. It would have been easy for Selwyn Lloyd to have introduced this reform just by publishing his own backroom estimates in 1961. Instead, we have had to come to it by a long round journey, through unnecessary Neddies, DEAs, seven years of too high an exchange rate, ten years of too low investment, some bitterness, lots of double talk, and many disappointments. That's what happened to British planning.

SOME conclusions are appropriate, even in an article about a word that may be rapidly deteriorating into the status of a nonsubject. In European politics during the 1960's, the word "planning" has meant three things:

(1) It is highly desirable that governments of modern industrial countries should let their people know what approximate economic growth rate they think they are aiming at in any particular period, and how it compares with the probable growth in the country's productive capacity. In one form or another, most modern countries

now in fact do this. Sometimes this is called "planning," and sometimes not.

(2) In a few modern countries without adequate capital markets, governments have tried to influence or arrange the directions in which new capital flows. This function of "planning" now seems to me to be everywhere in some retreat. Even the Russians and the East European communist states are trying to establish a sort of market for new capital provision, in place of a system of tight central direction. The French *commissariat du plan* is now less important than it was in the early 1960's, and the private French capital market is more important. In Britain, the "national plan"—which tried to discuss relative growth rates in a country with a fully developed capital market—became merely a joke.

(3) It is entirely sensible that governments should plan their own expenditure on cost-effective principles, and that they should classify central and local governments' uses of resources according to the ends rather than the means of government policy. One of the newest systems of classification, imported from the United States, is called in Europe "planning-programming-budgeting." There is therefore a tendency among social-democratic politicians in Europe just now to say that this is what they always meant by planning: namely, that governments should regulate their own spending according to the systems that the best private corporations use, but that the worst private corporations don't. In this sense, we can all say that we are in favor of planning, just as we all still say that we are against sin.

But, it is considered a bit old-fashioned to say either thing too loudly in Europe at the beginning of the 1970's. Planning, for the most part, is something that the statesmen of the last decade wrongly thought they did.

10

Japan's Galbraithian economy

MARTIN BRONFENBRENNER

JAPAN's remarkable economy has been growing, in recent years, at the rate of 10-12 per cent a year. At the present rate—if it can be maintained—Japan will be the second largest industrial power (based on GNP) in the 1980's, and conceivably could be the first in the year 2000. What is the "secret" of Japan's economic growth—a record which has now outstripped the Soviet Union, and which makes the 19th century history of Western capitalism look pale? The answer lies in a unique combination of Japanese industrial conglomerates—*zaibatsu*—tied to a centralized financial-investment structure, and a savings rate, accepted by a compliant population, which is the highest in the world. It is a remarkable combination of a "Protestant ethic" with a Galbraithian-type economy.

A Galbraithian economy (as per Professor J. K. Galbraith's *New Industrial State*) is one dominated by large corporations, many of them conglomerate. These corporations finance their own growth out of retained earnings from the public and from advantageous deals with affiliates and subsidiaries, including commercial and investment bankers. They can manipulate private and public demand for their products by advertising and lobbying, so as to sell profitably whatever they decide to produce. In this comparatively lower-risk environment, what is the ethical-economic justification for the profits of these corporate giants? The justification seems to be that relative risklessness makes it safe for the corporate management and technical staff—"technostructure" is Galbraith's term—to give technology a free rein, without worrying unduly about spoiling or outrunning one's markets. Private output and its growth rate can both rise and remain high. Private employment also remains both high and secure. The public sector, however, may be neglected, and the wrong goods may be produced, with reference to the quality of life.

I propose to discuss Japan as an exemplar of a Galbraithian economy. Galbraith himself is no "Japan hand," nor does he play any direct role in Japanese economic guidance. The paradox, however, is that the "modern" or "big business" sector of Japan's corporate economy comes closer to Galbraith's vision of the new industrial state than the American economy, and more than any other economy I know about.

The major Galbraithian feature of the Japanese scene is the development of the corporate conglomerate, or *zaibatsu*. Partly as a result of generations of back-scratching and special-dealing within corporate families and between corporate in-laws, the power of the three largest traditional *zaibatsus*[1] has reached the point that the Japanese city dweller can hardly live a day without patronizing at least one of the three. The contact may be through his bank, his toothpaste, his breakfast egg, or some household electrical appliance, but it is there. I have bought a building lot from an Osaka *zaibatsu*. A *zaibatsu* affiliate searched the title, fenced the land, and put in the streets and sewers which service the area. The same *zaibatsu* owns the bus and electric railway lines which reach the site from downtown Osaka. If I decide to build, a *zaibatsu* affiliate can design my house for me, and another can build it. Another affiliate owns a leading Osaka department store, located in the *zaibatsu's* electric railroad terminal. Another affiliate owns an amusement park outside Osaka, complete with dance company, along the same electric railway line. Yet another affiliate runs an Osaka big-league baseball club, complete with stadium located conveniently to the *zaibatsu* electric railway line.

I have no idea what else "my" *zaibatsu* owns, but as

[1] These are Mitsubishi, Sumitomo, and Mitsui, in order of size. A fourth great prewar *zaibatsu*, Yasuda, has chosen to confine itself largely to finance and insurance since 1946.

zaibatsus go, this one is minor and provincial, confining itself so largely to the Osaka-Kobe-Kyoto metropolitan area of less than ten million souls. I have no idea whether those who have sought to put together American conglomerates had read the business history of any of the great *zaibatsu*, or were simply following the "logic of the economic situation," but clearly in that respect, the Japanese have long led the way.

IT would be a mistake to think of Japanese big business as limited to affiliates of pre-war *zaibatsus*. Many, if not most of the leaders in the new, "hot," technical fields arose quite independently of *zaibatsu* ties. (I should not be surprised to see some of them founding 21st-century *zaibatsus* of their own!) Such examples of independents, without benefit of *zaibatsu* corporate cousins, include such prestigious and innovative concerns as Hitachi, Honda, Matsushita, Sony, and Toyota. Most of them concentrate on postwar technology, as applied to electronics or to transportation.

The conflict between new and old firms—future and present *zaibatsus*—has had an interesting, unexpected, and un-Japanese consequence on the executive labor market. Lacking any sufficient lower- or middle-level executive cadre, many booming new companies have raided the older ones for executive talent. Japanese workers, both blue- and white-collar, used to be "Sumitomo Electric" or "Mitsubishi Heavy Industry" men for life, never thinking of changing jobs and rising more or less uniformly, each along his own company escalator of ranks. Now there is the danger that certain overlooked young geniuses will jump to newer companies and gain several steps on the escalator. The older companies are beginning, like American counterparts, to make hard decisions rather early, separating sheep from goats in each year's entering class of aspiring junior executives. The fast sheep must be moved ahead at an un-Japanese rate of speed, or some upstart concern will grab them. The slow goats are consigned to escalator-riding, pre-war fashion. I do not know what this change will do to the rates of suicide, nervous breakdowns, and stomach ulcers in the Japanese executive ranks. My guess is that the effects will be minimal, for no goats will be fired outright. Japanese business has developed a genius for disguising executive goats long enough to get those goats tranquillized and reconciled to goatishness.

WHILE corporations follow the Galbraithian style, the Japanese economy as a whole was never modelled upon *The New Industrial State*. The Galbraithian pattern is therefore modified somewhat, particularly as to finance and to the role of government. In the Galbraithian model the corporate giants finance most of their own growth by the internal investment of retained earnings, and raise the rest by selling shares to the public. Japanese giants, however, pay generous dividends to present stockholders, and finance their growth largely by borrowing from commercial banks. Each company's credit and expansion funds comes from one, or more commonly two or three, of Japan's gigantic nation-wide branch banks, called city banks (*tokai ginko*). In *zaibatsu* cases particularly, there is commonality of ownership between borrowing firm and its primary lending bank, with *zaibatsu* banks and their officers investing heavily in the stock of their *zaibatsu* affiliates. The Mitsubishi Bank, for example, is an important stockholder in most Mitsubishi companies, the Sumitomo Bank in most Sumitomo companies and so on. (The reverse pattern of ownership, of banks by companies, is less important.)

The credits from Japanese commercial banks to Japanese companies take the form of short term loans, as in America. These loans are constantly renewed and expanded, however, so that a Japanese bank's loans often amount to over 80 per cent of its total deposits, even including deposits created as a result of the lending process.[2] If this line of credit were ever cut off when the borrower was in trouble, instead of being expanded for rescue operations, the borrower would almost certainly fail. But the lending bank would be in trouble too, because its previous loans to its failing borrower would become frozen assets. So constant credit expansion by banks to their major customers insures firms and banks simultaneously.[3]

The borrowing companies' demands for finance are limited by the interest charged by the banks. Another check to credit expansion is the banks' own resources.

---

[2] The ratio sometimes went over 100 per cent in the 1950's. During "overloan" episodes, the city banks were lending out their total deposits plus their own borrowings from the Bank of Japan!

[3] Community of interest with their borrowers sometimes takes banks further than this. The city banks help support the price of their major borrowers' shares on Kabuto-cho, Tokyo's Wall Street, as a matter of further mutual protection.

These can, however, be supplemented by the banks' short-term borrowings from the Bank of Japan, which plays much the same central banking role as the American Federal Reserve System. These borrowings have also been renewed and expanded over the years, to the extent that failure of any city bank might compromise the Bank of Japan itself!

So we have the Japanese financial tripod: the big business firm operating with other people's money; the nation-wide city banks, tied to their major borrowers; the Bank of Japan, tied to the city banks. Each leg of the tripod supports both the others. None can fail without involving both the others. Japan's big companies are safe, not so much because of any diabolical control over demand—a Galbraithian explanation for America—but because the banking system cannot let them fail. The results of their safety in Japan in terms of measured output, measured growth, measured employment, and measured technical progress, is much the same as the results in *The New Industrial State.*

The Japanese financial system has two sets of victims, however. One is the unaffiliated small business man; another is the fixed-income consumer. The unaffiliated small business man is by definition neither a favored subcontractor, a favored supplier, or a favored sales outlet for any of the big companies, *zaibatsu* or otherwise. He suffers because it is his line of credit which is fragile, and sensitive to the slightest hint of credit rationing or tight money. In any recession, he is bankrupted by the thousands, while his workers lose their jobs and seniority.

The fixed-income consumer suffers because the constant expansion of bank credit pushes up the money supply, which has pushed up the price level in its turn. Some recent figures illustrate this point. The Japanese money supply was less than 3,000 billion yen in January 1960. It was five times as much, or 15,000 billion yen, in June 1969. Since the autumn of 1963, the annual rate of increase has been slightly more than 15 per cent. Consumer prices, with the 1965 level taken as 100, had risen to 120 in the four years ending June 1969, without the impetus of a Vietnam War or a swollen public budget. (The fixed-income group is, however, less important in Japan than in most Western countries; most income receivers have at least kept up with the inflation.)

In *The New Industrial State*, the crucial economic function of government is to generate a profitable market for the most advanced and sophisticated products at the frontier of the system's technology. If the frontier is supported, it will pull the rest of the technology after itself. This public function Galbraith assigns primarily to military programs, with important assistance from paramilitary ones like atomic energy and aerospace.

This is not quite right for a semi-disarmed Japan with a defense budget below 1 per cent of GNP. Here the Ministry of International Trade and Industry (*Tsusansho* to the Japanese, MITI to the foreign community) has taken on a wide range of extra-legal expansive, regulatory, and protective functions. These are exercised in the common interest of Japanese big business. This is why some commentators see "Japan, Inc." as the world's biggest combine. We will discuss three or four interesting and important examples.

1. One MITI function has been the use of controls over corporate licensing, imports, and conversion of earnings into foreign currency, so as to keep foreign firms out of the Japanese market until Japanese ones grow big and strong enough to compete with them on at least even terms. If a foreign company establishes a wholly-owned subsidiary in Japan, or buys a controlling interest in an existing Japanese company without MITI's blessing, it finds itself unable to import whatever parts and goods-in-process it needs, or to repatriate not only profits but interest on its Japanese investment. As we have said, MITI licensing is not forthcoming if Japanese firms fear foreign competition. What comes instead is an exasperating and interminable series of queries, investigations, and delays, until the exhausted foreigner takes the hint and withdraws.

2. Another MITI function in aid of domestic big business has been the elimination of excess competition[4] by encouraging mergers and cartels, which raise the size of Japanese firms. The theory here is that, without the pressure of such competition, the "chosen instruments" select-

[4] A rough definition of excess competition (*kato-kyoso*) includes any form of competition which results in holding firms below their optimum size, or which guts their profit margins greatly for long periods, or which forces them to behave in non-Japanese ways to keep their heads above water. "Non-Japanese behavior" in turn includes the laying off of permanent staff, stopping payments to workers and subcontractors, and eliminating the profit margins of affiliated suppliers and sales outlets.

ed for survival by MITI will grow faster, and attain earlier some optimum size at which they become competitive with foreign, and particularly American, rivals. An example has been the Japanese automobile industry, where MITI and the automobile companies, working in cooperation, reduced a "middle-sized nine" to a "big two-and-a-half" within approximately a five-year period.

3. MITI must approve technical assistance contracts involving payments by Japanese firms to foreigners, in connection with the application of advanced foreign technology. This power is exercised not so much to keep the payments down and protect Japanese companies from exploitation than to prevent any one Japanese corporation from opening up a technological gap and making domestic rivals obsolete. Thus Japanese firm A will not be allowed to make payments to I.B.M. under a contract involving exclusive rights to some aspect of computer technology until Japanese firms B and C have signed equally beneficial contracts for exclusive rights to equivalent techniques from General Electric, Sperry-Rand, Siemens, or Olivetti. Monopoly is prevented by this delay. At the same time, nobody's capital need be written off, and nobody need take capital losses, because of some rival's superior contacts overseas.

4. MITI regulates the fine structure of Japanese import regulations, within the limitations of broad general laws. Some of its changes become effective immediately on promulgation, before they are formally published. MITI interpretations and changes are made with a protective bias, to reduce foreign competition to an extent greater than implied in the text of increasingly liberal general laws. For example, when sales of used foreign cars were injuring Japan's automobile industry in the late 1950's, MITI used the forgery of import-duty receipts on certain used cars as an excuse for banning used car imports indefinitely. Two later examples of similar tactics: foreign radio companies have been allowed to import only the parts for primitive crystal sets, while parts for modern sets must be bought from Japanese competitors; automobile imports have only been permitted if the machine is fully assembled, although import of parts for assembly in Japan requires much less shipping space and is therefore much more economical.

LESS Galbraithian, but important for understanding the Japanese economy, are certain elements of economic sociology. Some of these factors operate within individual firms; others relate to the behavior of workers. We can do little more than mention four such factors here.

1. In Japanese firms, more people are involved in decision-making than is the case in the United States, where one-man rule is a common alternative. In the Japanese collective-decision process, an attempt is made to provide some advantage for every participant, and prevent loss of face for any of them. Some Japanologists attribute this system to the more general style which derives from the density of Japanese population, the low level of mobility, and the consequent need to "live together" for long periods. A disadvantage of the Japanese way is the slowness of the decision process; this is a standard complaint of the foreign business man in Tokyo, particularly when the short-time visitor sets a high value on his time. The offsetting advantage is a greater degree of willing cooperation in decisions once they are made, and smaller problems of foot-dragging and sabotage. At the same time, the firm as a whole is willing to risk more because no individual can be held to account for any failure. The Japanese openness to technological progress certainly suggests that the Japanese way, for all its short-term slowness, has its long-term advantages.

2. The high degree of paternalism. In order to train Japanese workers and executives in Western industrial practices, without risk of losing the trained men to their rivals, late 19th century Japanese firms embarked on an elaborate system of lifetime employment guarantees and paternalistic fringe benefits. The most important of these benefits, low-cost company housing (housing has been in short supply ever since the bombings of World War II) family allowances and severance pay (upon retirement). Naturally, the fringes add to labor cost. So far, however, they seem to have paid for themselves in higher morale and productivity. (It is probably a mistake to trace the Japanese system to pre-industrial feudalism, as used to be the common practice.)

3. The Japanese worker has employment security, and neither craft nor trade unionism has taken root in the plant. Japanese workers and unions are therefore almost entirely unconcerned with job boundaries and jurisdictional disputes. The Japanese worker is less worried than the American worker about losing his obsolete job to automation or other forms of progress, when another job is

waiting for him at equal pay in the same company, and usually in the same factory. If we compare Japan and America, using the standard criteria of morale, productivity, and the speed of technical progress, the advantage is almost entirely on the Japanese side.

4. Investment and growth are, in the last analysis, financed by saving. This saving may be voluntary, on the part of individuals or corporations. It may be the "forced saving" of inflation, which reduces consumption. Japan's high propensity to save, in which the wage-earner shares, has been the wonder of other countries with similar levels of income per head. During the recent boom the saving ratio has risen from 24.3 per cent of GNP in 1952-54 to 37.7 per cent in 1966-68.[5] (The higher figure is comparable with those squeezed out of the USSR by Stalin during the first Soviet Five Year Plan!)

I cannot explain this peculiarity of the Japanese saver, or estimate how long it will hold up under the mounting pressure of advertising. Two factors making for high savings, however, are the financial system and the method of wage payment. The financial system concentrates credit, as we have seen, on industrial expansion. There is little left over for the consumer. If a worker or a small business man wants a large durable item—a house, a car, a building lot—his down payment will be much larger proportionately than in America, and he will have to save it in advance. The wage payment system has come to feature two substantial bonuses (several months' pay), the smaller one at mid-summer and the larger at year-end. (This device developed during the 1945-49 inflation, when the bonuses were necessary for subsistence because prices rose so quickly.) Bonus income, being uncertain in amount and coming only twice a year, is distinctly more likely to be saved than normal income. In fact, it is an unusual practice for Japanese workers, despite their reputation for frugality, to live on their ordinary monthly incomes. The

---

[5] Professor Hugh Patrick of Yale has culled these figures from Japanese Economic Planning Agency publications. They include corporate and public saving along with personal saving. The personal saving component more than doubled (from 6.0 to 12.3 per cent of GNP) in the periods compared. Professor Patrick's analysis, entitled "Fiscal-Monetary Policy and Economic Growth," was published in English in the Japanese journal *Keizai-Kenkyu* for May 1970.

common practice is for them to go slightly in debt in ordinary months, and make up their arrears (with a good deal left over) from the next bonus.

SINCE 1960 the Japanese economy has indeed been the wonder of the world, exceeding even West Germany and surprising its own official Economic Planning Agency, whose targets have regularly proved conservative when compared to actual performance.

One persistent question has been: how long can the great growth last? Growth needs raw material; for Japan, raw material must be predominantly imported. How long can Japan's exports pay for these imports, against worldwide competition? Growth needs labor as well. How long can Japan provide it, especially for ill-paid small industry, with abortion cutting down population growth, and traditional rice-paddy labor reserves nearing exhaustion? Growth needs savings—individual, corporate, and public savings—invested constructively and domestically, in what economists call "directly productive facilities." How long will Japan supply enough savings and directly productive domestic investments, against the counterpulls of the country's private living standard and relatively neglected social capital? Growth needs technical progress. Once caught up technically with Europe and America over the entire range of industry, how long can Japan continue a progress rate so far above theirs? Finally, growth has involved, at least for Japan, a shifting industrial structure, with investment diverted to those industrial sectors with the highest, or fastest rising proportions of value added to raw materials and capital depreciation.[6] How long can the Japanese industrial structure continue to be twisted in these directions without eventually settling down, on the basis of "this far, and (almost) no further?"

Foreign doubters and pettifoggers about Japanese growth (this writer included) were seldom sought out eagerly in the Japan of the 1960's. Timorous suggestions smacking of market economics, conventional wisdom, or standard foreign practice, have been met with polite queries like "And how much did *your* country grow last year?" But in the 1970's, more and more Japanese seem to be joining the ranks of doubters and pettifoggers. National pride in "income doubling" (*shotoku-baizo*) is giv-

[6] This point is stressed by Professor Shigeto Tsuru of Hitotsubashi University (Visiting Professor, Harvard University, Summer 1970).

ing place to concern about "price doubling" (*bukka-baizo*) or even "the anguish of a growth economy" (*seicho-keizai no kuno*). Along with the highest measured growth rate in the world, Japan has achieved, by globe-circling astronauts' reports, the most polluted atmosphere in the world. The atmosphere serves as a symbol for social problems neglected by "economic animals." For a number of years (culminating in 1967-69), Japan has also boasted the world's most powerful and violent New Left protest and student movements, if one excepts the short-lived French boil-over of May, 1968. Indeed, Japan's *Zengakuren* and *geba-gakusei* became temporary international symbols.[7] Symbols of what? Primarily, or so it seems to me, of protest against measured growth rates as indices of either genuine progress or genuine welfare.

There is no need to join the crowd of economists and statisticians predicting the year's, or the decade's, measured Japanese growth rate to the *n*th decimal place. Others can make more accurate calculations than I, on the basis of fuller and more recent data. Some recent reports (as of June-July 1970) indicate at least a pause in the real Japanese growth rate for 1970, as compared with recent years—9 to 10 per cent for the first half, 5 or 6 per cent for the second half, 7 or 8 per cent for the year.[8] (The precise figures, to my way of thinking, will not matter very much.) If this pause ends an era of restructuring Japan's economy toward higher-productivity and growing-productivity industries, which is the main thrust of Tsuru's explanation of the year's events, it may even be one of those pauses which "come to stay."

It seems rational, long pause or short, to conclude on a note of antigrowthmanship (stagnationmanship?) with the suggestion that several points—high measured growth, high atmospheric pollution, widespread protest against the anguish of growth—are related more closely than complacent Japanese growth-men have thus far cared to admit. This implies that measured growth rates, even the impres-

---

[7] *Zengakuren* is an abbreviation for the lengthy title of the All-Japan Union of Student Self-Government Associations. Sponsored originally by the American Occupation, it was patterned after the National Student Association in the United States! *Geba-gakusei* are students (*gakusei*) enamored of physical power for its own sake, *geba* being a Japanese abbreviation of the German *Gewalt* (power).

[8] These estimates I owe to Professor Tsuru's reading of the Japanese financial press.

sive Japanese ones, are far from a universal solvent for the discontents of industrial civilization. Perhaps they even light matches to the oil they cast on troubled waters and at faster rates than they cast the oil.

11

Sweden: some unanswered questions

ELI GINZBERG

THE stay-at-home American public knows Sweden as a country with a high suicide rate—a demographic tidbit given wide circulation by the late President Eisenhower; as a producer of movies that reveal more than they hide; and as the home of smorgasbord. The Swedes do have a high suicide rate—but it is probably no higher than our true rate, which social scientists estimate to be about double our reported rate. While they do produce many sexy movies, the "Romeo and Juliet" which is currently making the rounds of the Swedish pornographic circuit was produced in Hollywood. As for smorgasbord, it is put on the menu of a good Stockholm restaurant only with the advent of the tourist season. But even if the popular beliefs about suicide, cinema, and smorgasbord were correct, they would not provide an understanding of the most successful economy outside the United States or of a society that has been led by a Social Democratic government for the last three decades and which has neither urban slums nor rural poverty.

To understand Sweden requires attention to fundamentals. First, it covers a considerable area—it is about the size of California—but its population is no larger than that of New York City. Almost all of its people, about 85 per cent, are concentrated in the southern quarter; the forbidding north, with 25 per cent of the land area, has only 3 per cent of the people.

Geography creates climate, in this case rather forbidding. Even in the more benign south, there are only a few hours of sun and light throughout more than half the year. Small wonder that almost all factories and offices close down during July, when Swedes take off to the lakes and the countryside to bask in the summer sun.

Geography also creates, in this case, a certain degree of isolation. Sweden lies off the beaten track. Amsterdam, Brussels, Copenhagen are major stops for the traveler by road or air. But Stockholm, which is off by itself, leads only to the Arctic north, the land of the Laplander and the reindeer.

So much for the influence of geography. A word about history and economics. Sweden fought its last war in 1814, and while some scholars believe that wars accelerate economic development, no one questions that Sweden's prosperity was speeded by her neutrality during World Wars I and II. In fact, her Nordic cousins are critical of her behavior, believing that she was more concerned with material well-being than with national honor. The other aspect of the Swedish past that must be put into place is her relatively late modernization. Her transformation into a modern industrial state took place in the last quarter of the nineteenth century and her democratic institutions are even more recent. It was only in the present century that employers were prohibited from beating their workers and all citizens were granted the franchise. If there is a Swedish miracle, it occurred in the twentieth century.

The "Swedish way of life" also contains the following elements: (1) The long-established and long-entrenched feudal system, with its rigid class structure, was not eliminated by revolution; it was politically transformed. One consequence has been that class differentiations have continued although social mobility has increased. (2) The homogeneity of the population, in terms of cultural identification, language, and religious persuasion, is striking. Except for the recent large immigration from Norway and Finland, and the smaller inflow of political refugees and imported laborers, Sweden has no minorities except for tiny numbers of Laplanders and gypsies. (3) Finally, Sweden is intimately tied to the world markets. It literally must export or die—and this fact has been imbued in the public's consciousness and helps explain many of its actions and reactions.

DURING the last thirty years, Sweden has built a large number of economic and social institutions which have had a twofold aim: to assure a steady and high rate of economic growth, and to provide generous social benefits and welfare services to the entire population.

With respect to economic policy, the Swedes can claim with justice that they were the first practicing Keynesians—and, in fact, that their economists, particularly Wicksell, were ahead of Keynes. They can also point to having taken the lead in the design and implementation of an active labor market policy which recognizes the need for large-scale government training and retraining efforts as a necessary precondition for full employment. Until the unexpected strikes of last year, it seemed that the Swedes had also discovered the key to collective bargaining without work stoppages. No matter what the future holds, the success of Swedish industrial relations over more than three decades warrants admiration.

On the social front, Sweden was less the innovator than the pace setter. She has provided more social services than any other country for the widest possible range of contingencies, from conventional types of social insurance—unemployment, old age, illness, and disability—to such elaborations as paid vacations for housewives, government purchase of homes for workers who want to relocate away from labor surplus areas, and the provision of "archival employment" for workers who cannot stand the pace of the competitive market.

In Sweden, Social Democrats lead the government; sophisticated economic theories are the foundation for a high employment, rapid growth policy; strong egalitarian trends are reflected in high taxes on the rich and high social benefits for the less affluent and the handicapped. What more can one ask for in an imperfect world?

If the object of this exercise were to weigh Sweden in the scale of nations, the verdict would be easy: she has done well; she is doing well; and she is likely to continue to do well in comparison with the other major industrial nations, large and small. However, we are concerned here with juxtaposing her achievements, which are well known, with her emerging problems, which are only now beginning to surface.

SWEDEN is one of the most thoroughly organized societies in the world. Everybody belongs to many different organizations. Membership in an organization in Sweden means that the individual looks to the organization to make important and binding decisions, which he then follows. Americans also belong to many organizations, but for the most part their membership is for social or civic purposes. But even if economic memberships are compulsory, American employers and employees do not expect or permit their trade associations or their trade unions to make critical and binding decisions that shape an individual's destiny.

The Swedes acknowledge that they are enmeshed in a web of institutions but they are quick to add that the system works: the economy prospers; the society is attentive to welfare. The outsider, however, may raise questions. Those now in leadership positions have played a major role in creating many of the present institutions and in getting them to work. When I asked Arne Geijer, the President of the Swedish Trade Union Confederation, whether he is satisfied with the Swedish Labor Market Board, he said that he could hardly be dissatisfied because most of the ideas on which the Board operates originated in his shop. Many leaders are similarly gratified. But what of the younger generation who find these institutions in place, who know nothing of the struggles that preceded their establishment, and who played no part in bringing them into existence?

When the iron miners in the north found that their local union was repeatedly unresponsive to their complaints they broke with tradition and went out on an unauthorized strike which lasted several months. The facts that their employer was the government, that they were near the top of the wage scale, that the nation's balance of payments could ill afford the loss of ore exports, that "responsible unionism" required respect for the nonstrike clause in their contract—all these arguments were unavailing in the face of the unresponsiveness of management and the union to their grievances. Some observers believe that the iron miners' strike was an exception to the usually pacific Swedish labor relations. They even count as an exception the strike at the Volvo plant. But other students of the Swedish scene are less certain about this, and wonder whether the future will be more exceptional than not.

When we ask how this over-organized society succeeds

in functioning, the answer is readily apparent. The leadership group is relatively small and many ties hold them together. There is, then, a small and powerful Establishment. The Minister of Justice is the former attorney for the white collar union. The president of that same union has just been tapped to head the TV and radio monopoly. Key professors sit on the boards of the major companies. The trade unions and the government have overlapping leadership. The construction unions work both sides of the street: they are also major builders. University graduates are helped on their way to the top by their teachers who sponsor them, and in time they call on their teachers to act as consultants. The members of the same small leadership group meet each other in a great many different decision-making settings. They get to know each other well and they understand the need for amicable relationships. When problems arise, they are usually settled between friends on the phone and only later is the agreement committed to paper. And this male Establishment is strengthened and reinforced by ties of blood and friendship on the distaff side.

Query: How long will the new generation of young people accept the Establishment and permit it to continue to make all of the critical decisions? This question is not easy to answer. Even if answers were forthcoming, it would be difficult to know whether they will prove to be right. Yet this much appears clear: the more education a people acquires, the more affluent it becomes, the more it has access to a free press and other communication media, the less likely it will follow without criticism or challenge the policies and programs developed by its leaders. Here is a second reason to wonder whether this highly organized society may not soon face more of a challenge than an occasional wildcat strike.

If future challenges to the Swedish Establishment will probably result from nascent dissatisfaction on the part of the young, let us consider them. Sweden has long had and continues to have a "drug problem." One hears conflicting reports about the numbers who are involved and the extent to which they use addictive drugs. But there is sufficient restiveness among the educated laity to suggest that the use of drugs is widespread among the young.

What are the attitudes and behavior of youth toward such values as sex, work, and religion? Once again, hard facts are difficult to come by, but a few tentative generalizations can be ventured. Once one allows for the generally high average of education, income, and urbanization, it is not surprising to find a relatively accepting attitude toward premarital sex, reinforced by the ease with which contraceptive devices can be purchased through vending machines. As one Swedish friend put it, "Sex in Sweden is no different from sex in the United States except that we don't have any guilt about it." This much is certain: There is much less touching, fondling, kissing among Swedish couples in public than among American couples. Since the sauna is an established institution, mixed bathing in the nude is widespread.

A more subtle aspect of heterosexual relations is the new concern, at least among an aggressive minority, with women's rights. For all of Sweden's touted liberalism and socialism, Swedish women are not treated equally—any more than women are in the United States, the Soviet Union, or any other industrialized country. It is likely that the steadily growing efforts with respect to women's rights may in the long run contribute substantially to a recasting of many established institutions.

One indication of what lies ahead can be found in the realm of work. Many employers complain about the rising curve of absenteeism, and of resistance to overtime work, which is new among a population that has long been imbued with a strong instinct of workmanship. The employers frequently overlook the fact that, with more and more young married women in the labor force, and with marked shortages of child care facilities, it is inevitable that many young women will be forced to miss a day's work and that many young husbands prefer to spend their free hours with their wives and children than to work overtime. The old order is changing in Sweden as elsewhere, once people are given effective options between more income and more free time.

DESPITE its enlightened economic and social policies, Sweden has been slow to modernize its educational system, and in particular to broaden access to the university, even though an academic degree is generally a prerequisite for entrée into preferred jobs and careers. Nowhere does the class bias of Swedish society show more clearly than in the family backgrounds of university graduates. The constraints until now have been cultural, institutional, economic. The sons and daughters of blue collar workers did not consider the university within their purview. The *gymnasium* (i.e., the academic high school, on the German model), until now the only route to the university, had a limited number of places and the curriculum was heavily weighted towards traditional subjects. And while tuition at the university is free, and students receive a small government stipend, most of those from lower-income homes must take out sizable loans to cover their living expenses away from home.

Educational reform is now the watchword. The Swedes have opted for a comprehensive secondary school with three tracks—academic, technical, vocational—each of which can lead to the university. The number of university places is being substantially expanded. Yet a general uneasiness prevails. Even with the—by American standards—restricted numbers, many graduates are unable to obtain desirable positions. The fear is that this problem will worsen as the number of university graduates increases. In the United States, a university graduate is willing to contemplate taking over his father's motel or gas station. In Sweden, as in other European countries, this is to think the unthinkable.

Sweden has recognized the importance of reforming its system of higher education in order to increase occupational mobility. But responsible planners are concerned about the possibility that larger public and private expenditures for university education may be reflected, not in increased economic productivity, but in greater social tension. It is not easy to reconcile a strong drive toward egalitarianism with a continuing respect for the elitist quality of higher education. If good jobs depend on higher education and if higher education is both costly and difficult, then more equality can be achieved with respect to access—but not necessarily with regard to outcome.

THE trade union leadership is consciously developing policies and programs aimed at narrowing the gap between classes, and also between high and low-paid workers within the working class. It has used its influence both in collective bargaining and in the design of public tax and expenditure policy toward its avowed goal for some time. The recent revision of the tax laws to permit husband and wife to report their incomes separately, and thereby to take advantage of a lower rate, has improved the position of families with two earners. Nevertheless, the leadership may confront difficulties as it moves to restrict the gains of workers at the top of the scale while accelerating the improvement of those at the bottom. (The wildcat strikes in the iron mines in the north and in the automotive plants in the south were started by workers who were at the top, not at the bottom, of the wage scale.) The record of the last year is not encouraging. As long as there is a significant minority of professional men and businessmen who earn 100,000 Kroner ($20,000) or more a year, a worker who clears Kr. 40,000 does not consider himself well off; nor does he agree that his wages should be frozen so that workers in the Kr. 20,000 to 30,000 range can have an opportunity to improve their standard of living. An American family with an income of $10,000 is more aware of the things it doesn't have and cannot purchase—things that are within the reach of families with incomes of between $15,000 and $20,000—than of the fact that it is better off than a family with an income of $7,500. Swedish families are not significantly different from American in this respect. And it is worth noting that the distribution of private income in Sweden today is not much different from that in the United States—in both countries, the top fifth and the bottom fifth get about the same proportions of the national income.

One consequence of the recent focus on wage differentials in Sweden has been to call attention to a group at the bottom who have no income or formal means of support. A preliminary estimate released by the Commission on Low Income Investigation puts this group at 2.5 per cent of the population. In addition to this "poverty group," the Commission found that about 45 per cent of all persons who worked last year (2 million out of a total of 4.5 million) had an income which was only half the median of the regularly employed, full-time work force. (This median was $6,000.) The group who worked short time did so because of illness, unemployment, or for family or personal reasons.

The only way that Sweden has been able to stay out in front has been to keep its export industries competitive through improved organization, technology, labor. Smaller firms are continually being merged with larger ones, and even while new firms are formed and small ones grow larger, other firms both small and large are forced to close down. As a consequence of this ceaseless dynamism, many workers over 50 lose their jobs. While most of them get reabsorbed, a significant minority—perhaps as many as 1 in 5—do not. This problem has surfaced only recently but the authorities are concerned about the future because the

Swedish labor force has an increasing number of workers who will soon pass 50. The "sheltered work shops" which the government has expanded so that these redundant workers will be able to make at least a limited contribution to the nation's output are not an adequate answer to the problem of the older worker, partly because they are so costly to run.

Another unresolved problem where low income, unemployment, and redundancy intersect is presented by the north. The authorities have been making efforts to develop a regional policy for this area, where unemployment and underemployment run relatively high among the agricultural and forestry workers. Although nationalization has proceeded apace in agriculture, lumbering, and mining, the growth of new jobs has been insufficient to absorb both those who have been displaced and the new entries who are seeking employment. So far, the Labor Market Board has relocated large numbers of northerners before or after training. At the same time, other branches of government have pursued policies aimed at encouraging the location and expansion of industry in the north in an effort to prevent further depopulation. The two policies have not yet been reconciled. Nor have the costs of a regional development policy been added up. In fact, no careful estimate has been made of the prospect of success of such a policy even if it were supported by large-scale subsidies to private enterprise and large governmental investments. At present, the largest of the three urban communities in the north is Luleä, which has 60,000 inhabitants. The critical question is whether a city of that size, or smaller, can be transformed into a growth center. It may well be that the regional policy will founder on the critical issue of density.

THIS brings us to the last and probably the most tantalizing of Sweden's unsolved problems; how much scope does a modern society have to correct the imperfections of the market place, to improve the social environment, and to facilitate the development of human potential? The answer would seem to be: a great deal but not unlimited scope. By the time government extracts—as the Swedish government does—about two-fifths of the national income by way of direct and indirect taxes, it is likely to be creating disincentives to work. And in an economy such as Sweden's, which must export to grow, a further constraint is introduced: Sweden's costs, including her labor costs, must stay in line with those of her major competitors. There are no domestic palliatives that can correct an adverse competitive position in the international market.

We find, then, that Sweden's margins for maneuvering are restricted both because of taxes at home and competition abroad. What does this imply for the next stage in the evolution of this highly successful social democracy?

The answers depend, first and foremost, on the adaptability of the institutions to meet the challenges of the international market place while facilitating the realization of domestic objectives. In assessing the potential of the corporate and political structures to respond effectively to the pressures and opportunities that lie ahead, it would be well first to sum up the principal lessons that can be drawn from the experience of the last several decades.

The Swedish Social Democrats almost completely ignored the dogma that a "progressive" state should assume ownership and control of basic industry. With a few minor exceptions, such as starting a steel mill in the north as a regional recovery effort, the government has until recently left the entrepreneurial structure largely intact. The only important sector where a change occurred was in the construction of new homes, where government and nonprofit organizations increased their share of new output from around 10 per cent to over one-third. Stockholm's Enskildabank, in *Some Data About Sweden 1969-70,* stated that government ownership "is primarily limited to a few special sectors, such as iron ore mining, public utilities, and transportation, where government owned enterprises account for over half of the produced goods and services. In 1968, about 200,000 people were employed by government owned companies. This was 6.5 per cent of total business employment."

Thus, the private sector was not threatened by government takeovers and it was largely immune from threats to limit its freedom with regard to mergers and consolidations, pricing policy, and foreign investment. In fact, the government encouraged mergers in the belief that if Sweden were to continue to compete successfully in foreign markets—its exports account for about one-fifth of its output—it had to rationalize and remain in the forefront of the technological race.

The post-war trade-off between government and business created optimal freedom for business to improve its productivity and its profits while permitting government to obtain an ever larger share of the gains for public investment and consumption. Just prior to the outbreak of World War II, the central and local government share of GNP amounted to about 15 per cent; in 1969, it was more than double that figure. Private consumption declined from two-thirds to just over 50 per cent of the total. But since total consumption in current dollars increased twelvefold, this meant that (a) the consumer enjoyed large increases in disposable income while (b) government was able to provide a corresponding increase in public goods and services, and (c) private investment in this period increased its share from 10 per cent to 12.5 per cent of GNP. Between 1950 and 1967 expenditures on welfare programs increased over sixfold in current dollars; in 1967, these expenditures accounted for 17 per cent of net national income.

So much for a highly successful past. What about the future? We can begin by noting that the special advantages of a high demand in the international markets for Swedish products, and the absence of strong competition, both of which characterized the 1950's and which led to high corporate profits, are probably gone for good. For Sweden to hold its own, and particularly to improve its position, it must move into and strengthen its position in fields of advanced technology that require large organizations with access to large capital.

In a speech on March 24, 1970 before the Swedish Chamber of Commerce of the United States, Kristen Wickman, the Swedish Minister of Industry, sketched the steps that have been initiated and the direction of the changes that will be taken to enable Sweden to meet these new challenges. He called attention first to the newly created Investment Bank for Sweden which is in a preferred position to make large risk-taking investments. It has already helped to modernize the critically important paper and pulp and steel industries; and it has recently entered the international capital markets to assist Swedish industry in financing investments abroad. The Swedish government is exploring new ways of exchanging information with business so that, together, they will be in a better position to determine the best prospects for innovation in the production and marketing of "new products, methods, and systems." Then Minister Wickman tipped his hand about a major departure: a move to direct government equity participation. He believes that such action is desirable, even necessary, on the following grounds: to speed development in backward regions, to take advantage of situations where "social profits" may be substantial but private capital is loath to take the risk, and to seize those occasions when a profitable business emerges on the basis of public funds previously invested in research and development.

Such a program might work if Swedish society holds together. This is the critical question. Can the overt alliance between the elitist groups who have dominated business, government, the professions, and the trade unions be maintained? Will young people coming of age be willing to fit themselves into the elaborate organizational structures which they had no part in fashioning and in the leadership of which they have no say? What will happen if they balk? If income redistribution has gone about as far as the working population will tolerate before they disown their leaders, what will take its place in the drive toward equality? Finally, how does one reconcile the imperative for excellence in science, technology, and industry with a society that, after years of Social Democratic leadership, is becoming restive about the persistence of class distinctions?

These are the questions to which answers must be found. It does not denigrate economics to say that successful reconstruction of the business and corporate sector is much easier to foresee than the adjustments which would be required to enable a high-income, well-educated, nonreligious people to find the new goals that will give meaning and direction to their individual lives and institutions.

12

Monopoly capitalism and neo-Marxism

RAYMOND LUBITZ

THE revival of radicalism in America has brought with it a search for a theoretical explanation of what is wrong with American society. Inevitably, many radicals have turned back to Marxist models of capitalist development. But these older models, derived from the Victorian capitalism of Marx's time, have little relevance to contemporary capitalism. An updated Marxist model, with a direct bearing on contemporary American capitalism, and offering suggestive theoretical explanations of our social ills, would be a major intellectual achievement. The one serious effort to provide such an analysis is *Monopoly Capital* by Paul Sweezy and the late Paul Baran. [1] This book has become an intellectual force in its own right. It is perhaps the most influential work in economics among radicals and is worth some serious attention. Unfortunately, in this case the more serious the attention the more critical it has to be.

Because Marx's model was based on the competitive capitalism of mid-nineteenth century England, the assumption of competition was central to his analysis. In *Capital*

[1] Paul Baran and Paul Sweezy, *Monopoly Capital: An Essay on the American Economic and Social Order* (New York, 1966).

he argued that competition between firms would ensure—this is one of his theories of inevitable crisis—a fall in the rate of profit. But more than a century has passed, there has been no such crisis, and orthodox Marxists have found themselves in the position of arguing that capitalists were destroying themselves even while they were obviously doing quite well. Baran and Sweezy begin with a different premise: American capitalism, they say, is not competitive but rather "monopolistic," and they propose to refurbish Marx's analysis by concentrating on the role of the large corporation in the economy.

Baran and Sweezy link the large corporation to the central theme of their work, the "generation and absorption of surplus under conditions of monopoly capitalism" (p. 8). The use of the term "surplus" is of major ethical significance. If you want to argue that an economic system is based on "exploitation" then merely pointing to an unequal distribution of income or wealth is not enough. It is always possible that these inequalities might be shown, in some sense, to have been "earned." If certain incomes can be defined as "surplus," those who get the surplus can establish no *economic* right to it.

In any case, Baran and Sweezy define the surplus as the "difference between what a society produces and the cost of producing it" (p. 9). The size of the surplus is an index of an economy's productivity and is increased by technological progress. They further argue that conventional economics does not examine the question: how is this surplus distributed? In orthodox economic analysis, higher output (or increased GNP) shows up as *someone's* higher income. But no distinction is made between "necessary" and "surplus" income; nor is any social judgment made on how the fruits of technical progress (the *surplus* in the Baran-Sweezy terminology) are distributed in the society.

BARAN and Sweezy claim that the benefits of technical advance are distributed highly unequally in monopoly capitalist society—*and* that this unequal distribution is determined by the activities of the giant corporation. The surplus, therefore and unsurprisingly, falls mostly into the hands of the corporate elite, in the form of profits, leaving the majority of people nearly as poor as ever. More than that, using a Marxist language (but not always a Marxist methodology) they seek to revive the thesis, once popular in the late 1930's when it was forcefully stated in Keynesian terms, that the American economy is "inherently" stagnationist and that there is still an endemic crisis in the system based on this tendency to stagnation.

Crucial to the Baran-Sweezy thesis is the highly orthodox assumption—from which such economists as J. K. Galbraith, Robin Marris, Herbert Simon, William Baumol, and Carl Kaysen dissent—that corporations do aim to maximize profits, as well as pursue other corporate ambitions (maintaining one's market position, achieving a steady or modest rate of growth, etc.). Baran and Sweezy argue further that technological progress, which cuts production costs continuously over time, will—contrary to Marx's propositions—expand corporate profit margins and cause aggregate profits to increase as a share of the national product (p. 71). They then "provisionally equate" these aggregate profits with *the surplus* and formulate, as a law of monopoly capitalism, that "the surplus tends to rise both absolutely and relatively as the system develops" (p. 72). But consumption by individual capitalists and investment by the giant corporations can absorb only an ever smaller part of the surplus. Yet if the surplus is not absorbed, it will not be "realized," that is, resources will not be fully employed—and that is why the economy will show a tendency to stagnate. Therefore, the society has to turn to such activities as civilian government expenditure, military spending, and massive advertising; (the last absorbs resources unproductively and stimulates consumption). *Monopoly Capital* asserts that none of these will offset the stagnationist tendencies of the system, although irrational and wasteful forms of surplus absorption will proliferate.

The key concept of *Monopoly Capital*, "the surplus," suffers from ambiguities—which is not surprising because it is the offspring of an uneasy mating of Keynesian and Marxist concepts. Baran and Sweezy never explicate their definitions of the term, nor do they use the concept consistently. Rather the reader is referred to Baran's earlier book, *The Political Economy of Growth*, for an extended discussion of the concept. But in that book there are *three* different meanings of the surplus, and *Monopoly Capital* is silent as to which, if any, of these meanings it understands as "*the* surplus."

Perhaps the most useful way to achieve a meaningful and stable basis for understanding their definitions of "surplus" is to relate them to the more orthodox conception of national income. There are two ways of looking at national income, and both add up to the same total, i.e., the value of goods and services produced in the economy. The first is *national product,* which is the sum of all expenditures—on consumption, on investment, and by government. The other is *national income,* which is the total of the income *earned*—as wages, rents, profits, or interest—by all the "factors of production" contributing to the national product. Since all expenditure on product winds up as someone's income, national income is by definition equal to national product.

Now, the first and most direct definition of "the surplus" given in *Monopoly Capital* is "the difference between what a society produces and the costs of producing it" (p. 9). But later, "surplus" is "provisionally" defined as aggregate corporate profits (p. 72). In national income terminology the Baran-Sweezy "surplus" is close to "national income *minus* wages" (of those engaged in producing the income). (The concept is also close to Marx's "surplus value," whence it presumably derives.) This surplus is property income (profits, interest and rent), but following *Monopoly Capital* we can call it simply "profits"; national income thus equals wages *plus* "profits." This surplus may be directed toward different uses—capitalists' consumption and investment, government expenditure, and "waste" (e.g., advertising and salesmanship). If we had the data, we could calculate the "national economic surplus" by two different methods: either as profits received (the income-surplus) or as expenditures out of profits (the output-surplus). Because national income by definition equals national product, the two sums should give identical *measures* of the surplus. What we cannot do is *add* the two measures (or parts of them). Yet, unbelievably, this is just what Baran and Sweezy propose! They say that, in order to arrive at an "expanded" concept of surplus, "other forms" of the surplus besides property income "assume decisive importance" and that there is in addition to aggregate profits some other "fully developed concept of the economic surplus" (p. 72, note 2). What the rest of the text presents as other forms of *income* surplus are in fact *expenditures,* i.e., government expenditure and "waste." In effect, the authors have taken expenditures from the *product* side of the national income equation and added them on to the "surplus" on the *income* side.[2]

Baran and Sweezy might make a rejoinder along the

[2] That I have not misread them in saying that Baran and Sweezy have simply performed an illegitimate arithmetic operation (moving a term from one side of an equation to the other and adding rather

following lines. Adding government spending to profits to derive the surplus would be wrong if government spending in fact reduced profits. But it does not, because corporations can shift the taxes which finance the government spending. Thus, government spending is spending out of "surplus" because in its absence the resources would simply have been unemployed—"unrealized surplus" shows up as unemployment of capital and labor. The spending does not reduce the surplus available to the private (i.e., corporate) sector. Government spending can, therefore, legitimately be added to profits (after taxes) without double counting.

THERE are two things wrong with this reasoning which seems to be embedded in their model. To equate *all* government spending with surplus assumes that no government taxes come out of wages. This, as all of us who look at our paychecks know, is absurd. Secondly, even if one supposes that, in the absence of government spending there would be equivalent levels of unemployment, such spending still cannot be thought of as pure surplus. After all, additional workers are required to produce the additional output which the spending creates. If "surplus" is output less the cost of producing it, then government spending creates *non*surplus incomes (wages) and cannot be legitimately added to profits to arrive at an "expanded" surplus concept.[8] It might seem incomprehensible that Baran and Sweezy can include all government spending as surplus, when taxes on wages finance part of this spending (more, of course, than corporate taxes do). But they have, in fact, unintentionally made wages part of the surplus!

Baran and Sweezy's real trouble is that they want "the surplus" to represent two different magnitudes: the category of income derived from corporate profits (property income), and the class of nonsubsistence expenditures in an affluent society. In a subsistence economy we might, as

---

than subtracting it), is illustrated by the statistical appendix to the work. There the surplus is represented as total property income *plus* a series of expenditure items.

[8] The analytic device that *Monopoly Capital* constantly resorts to, equating unrealized surplus to unemployed resources, is based on a confusion of Keynesian unemployment and Marxian surplus value. It clearly derives from what Baran called the "potential surplus," an output concept, that could consistently include unemployed resources; however the total of unemployed resources cannot be regarded as income surplus.

a useful theoretical simplification, equate property income and "surplus," because wages will be directed primarily to subsistence goods and property income will approximately equal nonsubsistence expenditures. But the Baran-Sweezy theory of the corporate surplus, which is specifically designed to link the two magnitudes in an advanced society, is doomed to failure because property income is simply so much smaller than nonsubsistence expenditures.

HAVING created an illegitimate "surplus," Baran and Sweezy then assert that this surplus has a "strong and systematic tendency" to rise. To say that there is a tendency of the surplus to rise as a proportion of national income is equivalent to saying there is a tendency for the share of wages to fall—because wages plus surplus exhaust national income. (Notice, however, that the only proof offered for this tendency concerns *corporate profits,* and yet they insist that profits, as well as all property income, are only part of the surplus.) Their argument is that under oligopoly conditions, declining costs due to technological progress leads to everwidening profit margins: "the monopolistic structure of markets enables the corporations to appropriate the lion's share of the fruits of increasing productivity directly in the form of higher profits." Note that this is the crux of *Monopoly Capital's* attempt to prove that the fruits of technological progress accrue to "surplus" and, therefore, to the corporate elite.

Any theory of the tendency of profits to rise must show that real wages rise less fast than labor productivity. If real wages and productivity rise at the same rate, the division of national income between wages and profits will be constant. However, Baran and Sweezy provide no explanation at all of the determination of real wages; therefore the *statement of the model is itself incomplete.* There is thus no proof of the tendency of the surplus to rise.

Moreover, the historical data for the period of monopoly capitalism fails to show any sign whatsoever of a rising income surplus. It is clear from the record that the surplus, interpreted as the share of property income, has fallen in this century.[4] Employee compensation, the wage share in its simplest meaning, *increased* from 54 per cent of national income in 1899-1908 to 69 per cent in

[4] For a discussion of the difficult statistical problems involved in interpreting the historical record, and for much of the data used here, see Simon Kuznets, *Modern Economic Growth* (New Haven, 1966), ch. 4.

1954-60. However, to reveal fully the relative fortunes of capital and labor, the income of entrepreneurs and the self-employed must be separated into implicit wages and implicit property income, and allocated to the overall totals for wages and property income. Kuznets' estimates for the share of property income, after this adjustment, show that it *falls* for the period 1919-28 to 1954-60. Moreover, we can perform a direct test of the Baran-Sweezy argument, and compare the trends in real wages and labor productivity. The increase in labor compensation per man-hour in manufacturing rose over the period 1909-14 to 1955-57 by 3.8 times, while productivity in manufacturing rose by only 3.3 times.[5] Thus profits do not claim the lion's share of productivity's increase; in fact its share is falling.[6]

Baran and Sweezy might reply that it is a *tendency* of the surplus to rise—and that this tendency need not always be reflected by actual statistics. In response to Nicholas Kaldor, who argued that the share of profits has not risen, they do say that the increasing surplus is limited by the capitalists' ability to consume and invest it. A good part of this surplus is "unrealized" and therefore would not appear in the statistical record as an actual rise in the share of profits:

> Profits which are neither invested nor consumed are no profits at all. It may be legitimate to speak of the potential profits which would be reaped if there were more investment and capitalists' consumption, but such potential profits cannot be traced in the statistical record—or rather they leave their traces in the statistical record in the paradoxical form of unemployment and excess capacity (p. 76).

They go on to argue that there are, in the record, indicators of the rising *tendency* of profits. These indicators are

[5] The series I used are from the United States Department of Commerce, *Long Term Economic Growth, 1860–1965* (Washington, D.C., 1966), Series A165 (Kendrick) for productivity and B72 (Rees) for wages.

[6] We should mention for completeness that the structure of property income has changed, corporate profits rising, and other forms of property income falling, as proportions of total property income. Because of this shift, corporate profits as a percentage of national income are higher than 19th century levels, but the statistics on the share of corporate profits after taxes from 1929 show no increase at all. Also, the crucial Baran-Sweezy argument is that the share of corporate output going to profits has increased, and this is disproved by the productivity and wage trends.

the depression of the 1930's and "the persistent rise in the unemployment rate in recent years" which "lends strong support to the view that the problem of realizing surplus value is indeed more chronic than it was in Marx's time" (p. 75).

These are indeed remarkable arguments! An actually declining share of surplus does not disprove a rising tendency. The rising tendency *is,* however, "proved" by the *existence* of unemployment—in the 1930's and the late 1950's-early 1960's! But Baran and Sweezy do not show a long-run rise in the historical trend of unemployment and excess capacity. If neither realized profits nor unemployment rose, where is the evidence for a rising surplus?[7]

It is an old vulgar-Marxist game to "prove" nonobservable "tendencies" by taking *recessions* as *proof* of the tendency and business cycle *upswings* as *deviations* caused by temporary counteracting forces. Of the entire period of monopoly capitalism that Baran and Sweezy analyze, roughly 1870-1963, they discover problems of "surplus absorption" to exist in *less than one-third* of this time (1908-1915, 1929-39, and 1958-63). The rest of the periods were "exceptions." After a century of such Marxist methodology, a certain skepticism on the part of orthodox economists is understandable.

There are other pieces of data which are relevant for testing the basic Baran-Sweezy thesis. Their theory predicts an increase in profit margins. Profit margins are measured by profits per dollar of sales. This figure for all manufacturing corporations reached its post-war peak in 1950. At that time it was 7.1 per cent. In the prosperous years of 1965-68 it averaged 5.3 per cent. So there is not only no observable widening of the profit margins, there is an actual decline. Similarly, the rate of profit (on stockholder equity) reached its post-war peak in 1948 (16 per cent); in 1965-68 it was 12.6 per cent.[8]

It is interesting to illustrate the absurdities that easy invocations of "tendencies" can produce. Suppose the "surplus" does rise in step with productivity growth. The Kendrick series on labor productivity shows a six-fold growth of output per man-hour from the decade of the

---

7 Also Baran and Sweezy never justify their equation of unrealized surplus with unemployment, which I have argued is an invalid procedure.

8 *Economic Report of the President,* 1970, p. 262. I am not myself trying to resurrect the Marxist theory of the falling rate of profit. The post-war figures cited are intended only to disprove a rising rate. I do not believe in any "law" of rising or falling profit rates.

1870's to 1957. Suppose in the initial period output was 100, wages 50, and profits 50. A six-fold increase in output (from the given labor input) would make output 600, wages 50 (if real wages do not rise) and profits, or surplus, 550—or $^{11}/_{12}$ *of national income.* If profits do not actually rise to such fantastic heights, the Baran-Sweezy position (even allowing for some wage increases), is that *unemployment* would. If so, the six-fold increase in United States productivity over the last century, taken with the failure of the share of "realized surplus" to rise, would imply unprecedented levels of unemployment. But clearly this has not been the case.

THE Baran-Sweezy argument that the surplus has a tendency to rise is unsuccessful. However, it is possible to recast part of their book as a Keynesian-stagnationist tract with two major points: (a) that *without government intervention,* the capitalist economy will reach equilibrium at some level of substantial unemployment because of the weakness of the inducement to invest; and (b) that *for political reasons* inherent in a capitalist society the government will not undertake the policies necessary to bring about full employment.

To consider this modified version of Baran-Sweezy, we must briefly outline the Keynesian argument. In that scheme, national income equalled national output, but some of the income earned in producing that output will be saved rather than spent. To sustain a given level of national output, total spending must equal total output; this implies that desired *investment* must equal desired *savings* at that level. The Keynesian argument is that a sustainable output level may fall below the level of the full employment of resources because desired savings are not necessarily invested. Because unemployment may result, the Keynesian prescription is to increase government spending, without an equivalent increase in taxes (i.e., run a budget deficit), in order to stimulate demand and raise output and income.

In the 1930's a variant of Keynesian thinking (which had a basis in Keynes' own writing) stated that the capitalist system had tendencies toward long-run stagnation because of the weakening of investment opportunities. To achieve long-run full employment, it was said, government deficit-spending would have to grow as a fraction of output; this would be necessary to compensate for the growing gap between desired saving and desired investment. In *Monopoly Capital*, Baran and Sweezy are arguing the same stagnationist thesis in Marxist language.

Baran and Sweezy discuss the problem of absorbing the "surplus" as a problem of finding enough *demand* to purchase the aggregate output. In other words, "surplus" is used as a near-equivalent of the Keynesian *saving*, and the novel problem of "surplus absorption" is reduced to the familiar problem of finding investment opportunities and government-spending opportunities equal to desired levels of saving. Thus the Baran-Sweezy contention that the United States is increasingly prone to stagnation can be stated in a logically consistent form, as was the 1930's Keynesian-stagnationist view.

The prevailing opinion among economists today is that any tendencies towards weakness of aggregate demand can be offset by expansionary fiscal and monetary policy. Baran and Sweezy argue to the contrary, that profits generated by the corporate sector will be greater than the capitalists' needs for consumption and investment, and that government spending will not, for political reasons, be able to offset the weakness of private absorption. Let us examine each form of absorption—by capitalists and by government—to see if *Monopoly Capital* proves this thesis.

First, the capitalists. Baran and Sweezy argue that capitalists' consumption as a share of profits will fall because current dividends as a share of current profits (what financial analysts call the payout ratio) will fall. This is a fallacious argument derived from two reasonable assumptions: (a) corporations have a "target payout ratio" and (b) the target ratio will be based on an earlier year's profits. Given a "continuous rise" in profits, Baran and Sweezy manage to deduce that the actual payout ratio will fall continuously (p. 80). The deduction is simply false. Assuming a given growth rate in profits, and a given "target payout ratio," the actual payout *rate* itself will always be constant, although it will be lower than the target rate, and lower than it would have been if profits were steady.

Baran and Sweezy do not present any statistics on corporate payout ratios. In fact, the ratio appears to be rising: 1950—35.2 per cent, 1955-57—45 per cent, and in the 1960's—45.6 per cent.[9]

Capitalists' investment will also be sluggish, *Monopoly*

[9] *Economic Report of the President,* 1970, p. 260.

Capital argues, because whereas technological progress in a competitive system is a strong stimulus to investment, it does not have this role under monopoly capitalism. They argue that under *strict monopoly* conditions a monopolist will introduce new techniques more slowly than in competitive conditions (pp. 94-95); but they then go ahead blandly to apply the *strict monopoly* model to the real world of oligopolies. Earlier, and correctly, they use the oligopoly model to show how nonprice competition among oligopolists generates pressures to innovate. The monopoly argument is clearly the inapplicable one. (If it were applicable, moreover, what would happen to the tendency of surplus to rise?) Furthermore, no attempt is made to see if, in fact, the rate of productivity increase has fallen under the reign of "monopoly capitalism." If they had looked at the data, they might have seen evidence of a contrary tendency.

In the Baran and Sweezy vision, corporations are drowning in their profits and lack adequate investment outlets. They even argue (on the basis of the period 1953-62) that depreciation allowances are growing so fast compared to investment that "monopoly capitalism is increasingly able to take care of its investment needs from depreciation allowances" (p. 102). Baran and Sweezy do not explain why, if they are right on this, corporations do not increase their dividend payouts. Moreover, they argue, inconsistently, first that corporations adopt a payout policy which will cause dividends to fall as a ratio of profits and later that corporate profits are growing so fast that investment outlets are inadequate. They miss the point that corporate policy sets payout ratios to keep dividends at the minimum satisfactory to shareholders to allow corporations to exploit growth opportunities. If growth opportunities were shrinking then dividend payouts would rise.

A look at the data on internal corporate funds shows that such funds (defined as undistributed profits plus depreciation allowances) are smaller than total capital formation: in 1968 internal sources were $63.8 billion and investment $76.9 billion. There is no tendency observable in the data for internal funds to outgrow investment. (In 1957 the comparable figures were $30.6 and $34.7 billion.)[10] Recently, interest rates on corporate bonds have soared while companies bid for scarce investment funds.

---

[10] *Ibid.*, p. 264.

This reality is hard to square with the notion of a superfluity of corporate profits.

Outside of capitalist consumption and investment, *Monopoly Capital* treats the corporate sales effort and government spending as forms of surplus absorption. Baran and Sweezy emphasize that advertising not only "absorbs surplus" but also alters the division of income between consumption and saving. It is asserted that advertising exerts an "unmeasurable" but "probably very large" effect in increasing consumption. The thrust of the argument is that while much advertising is clearly wasteful and distasteful it is a necessity in a capitalist system in order to maintain demand. However, no evidence is actually presented that advertising does increase total consumption (as opposed to changing its composition), aside from some self-serving quotations from businessmen about the vital role of advertising in our dynamic capitalist society. The difficulties inherent in deciding this issue were well-explored in the Solow-Marris discussion in *The Public Interest*, No. 11, Spring 1968.

GOVERNMENT spending is the final element in the stagnationist picture. Government could, and in a rational society would, absorb ever growing amounts of the Baran-Sweezy surplus—but presumably the political and economic modalities in a monopoly capitalist society prevent this.[11] The oligarchy limits civilian spending despite great social needs because this spending will reduce the profitability of the private sector (the examples given are river valley development and public housing), or even when private enterprise is not threatened, because it will attack the structure of privilege enjoyed by the oligarchy (the example given here is education). Civilian expenditure which stimulates private enterprise (highways) is encouraged.

The easiest way to deal with this position is to look at some statistics. *Civilian* government expenditure as a percentage of GNP rose from 5.0 per cent in 1890 to 18.3 per cent in 1962.[12] Baran and Sweezy might say that

---

[11] *Monopoly Capital* is hard to follow on the limits to government spending. On pp. 149-150 they argue that the business community now accepts big government and strong pressures will keep government spending rising steadily, so the "big question" is "on what." But, much of the book, and particularly the chapters on government spending, argue that there are limits inherent in the system.

[12] Richard Musgrave, *Fiscal Systems* (New Haven, 1969), p. 94.

most of the increase in civilian spending came before World War II: *"given the power structure of United States monopoly capitalism,* the increase of civilian spending had about reached its outer limits by 1939" (their italics). This statement does not make clear whether the spending is limited as a share of GNP or in its absolute level. If it means the latter, it is clearly ludicrous. If Baran and Sweezy mean that government spending will not grow as a share of GNP, the data just cited show that it has. Baran and Sweezy also say that state and local expenditures will remain at a stable share of GNP. Yet they increased from $11 billion in 1946 to $121 billion in 1969; this ten-fold growth was far greater than the less than five-fold growth in GNP over the same years. In addition, not only has federal civilian spending grown enormously in recent years (from $17.2 billion in 1946 to $102 billion in 1969), but also social welfare items have grown greatly. Taking the three budget categories[13] that constitute the bulk of welfare spending, it turns out that they grew from $3.0 billion in fiscal year 1960 to $20.5 billion in fiscal year 1969.

The rigid model in which an all-powerful oligarchy prevents desirable social spending is obviously a fond illusion. To be sure, there are always strong forces opposing particular social policies. Interest group pressures do shape substantial portions of government programs while pressing needs go unmet. All of this is elementary. But an analysis of the forces supporting and opposing liberal social policies would, I am convinced, show that in some cases corporate interests acted the villain (tax legislation is a perfect example), but in other cases different forces constituted the opposition. *Monopoly Capital* lovingly details the horrors of urban travel and attributes them to the workings of corporate capitalism. But is the absence of efficient systems of urban transit really due to the opposition of the corporate elite? Wouldn't the urban centered corporate headquarters welcome them? (Are the neighborhood and community groups exempt as sources of opposition?) Federal aid to education was not opposed by the corporate oligarchy, maintaining its structure of privilege, but was rather enmeshed in the conflicts over integration in Southern schools and aid to religious education.

---

[13] These are "community development and housing," "education and manpower," and "health and welfare." *Economic Report of the President,* 1970, pp. 250-51. The data exclude social security programs.

MONOPOLY CAPITAL's discussion of militarism and imperialism contains some surprises. Military expenditure is required to maintain the American Empire, i.e., American trade and investment abroad.[14] But *Monopoly Capital* reverses the familiar Leninist analysis of imperialism which described overseas investment as an outlet for surplus capital. In the Leninist view, imperialism was an imperative of the capitalist system (the highest stage of capitalism) in which foreign investment offset the tendency of the falling rate of profit and prolonged the life of the system. This was a coherent (but not necessarily correct) theory. But the central theme of *Monopoly Capital,* the generation and absorption of the surplus, is logically unrelated to its discussion of imperialism. According to Baran and Sweezy, foreign investment *increases* the surplus rather than absorbing it. Imperialism is no longer the survival tactic of a doomed capitalist world. Rather the "theory," insofar as I can abstract it from *Monopoly Capital,* is that firms invest abroad because it is profitable to do so; also they will frequently wield political influence to secure their investments. So much is obvious. However, Baran and Sweezy also assert that American firms want monopolistic control over foreign sources of supply and foreign markets, and for this they require political domination enforced by American military power. The trouble with this argument is that most of the recent enormous growth of American investment abroad has been in Western Europe, where we deal with powerful independent states and cannot very easily extract economic favors; also, American affiliates in Western Europe are not generally the dominant firms in their industries and, therefore, do not exert the kind of market power they often have in the underdeveloped countries.

THERE is hardly a word in *Monopoly Capital* about the money supply, interest rates, or monetary policy. This is an astonishing omission. Theirs is an economics which has never outgrown the 1930's, and the Marxist concentration on "real" economic forces. The possibility that monetary policy can help stabilize the economy is never discussed; nor is the probability that some of the periods of stagna-

[14] In their single-minded determination to prove the severe limits on absorption possibilities, Baran and Sweezy say that there are constraints on military spending which inhere in the nature of modern weaponry and in the arms race. This will be good news to all of us who are appalled by the voracious appetite of the Pentagon.

tion Baran and Sweezy adduce as typifying deep-seated tendencies in the capitalist system were in fact, as Milton Friedman has argued, the result of mistaken government monetary policy.

This leads us into the chapter that is the capstone of the strictly economic part of the work, "On the History of Monopoly Capitalism." The book's model is "tested" against the economic history of the United States for the last 80-90 years. The surplus is presumed to rise throughout the period; its depressing effects become apparent in the absence of epochal technological innovations or wars. According to Baran and Sweezy, the railroad boom of the late 19th century absorbed the rising surplus, but the system showed its "true" character in the weak years of 1908-15. Now, most economic historians obviously accept the importance of railroad building (though some recent analyses by Fogel and others doubt this). The end of such a strong stimulus could indeed explain the weakness after 1907. A fall in capital formation of this magnitude would substantially weaken aggregate demand—but if the economy then recovers, as it did after an eight year period of lower-than-average activity, this proves nothing about *inherent* tendencies to stagnation. In the Baran-Sweezy methodology, downturns are caused by inherent forces and upturns are accidents. And they use even this procedure loosely. Thus, epochal innovations counteract the tendency towards stagnation—except when they do not. The automobile, also an epochal innovation, was not strong enough to forestall the Great Depression. The railroad did offset stagnation in the late 19th century. Why the difference? We are not given any explanation except that one was "strong enough" and the other "not strong enough" to avoid stagnation. This is circular theorizing on a grand scale.

For Baran and Sweezy, the Great Depression demonstrates the deep-seated forces of capitalism; it constitutes a proof of their theory:

> It may be said without fear of contradiction that the United States economics profession, nay bourgeois economics in general, quite literally has no theory to account for [this] phenomenon (p. 240).

Is it for this that Milton Friedman labored? Actually the work of Friedman and Anna Schwartz, with its mass of empirical verification, has convinced many economists that

Federal Reserve monetary policy (a severe contraction in the money supply) turned an ordinary business downturn into the Great Depression. But even if Friedman is wrong, the single *fact* of the Depression cannot prove the Baran-Sweezy theory. A theory should explain all—or at least most—of the facts to claim validity. Thus the period of the late 1950's and early 1960's is taken by Baran and Sweezy as further evidence of the inherent tendency towards stagnation. Nothing is said about the restrictive monetary and fiscal policy of those years. Yet in the mid-1960's the economy clearly turned around—without any epochal innovations and before any significant military escalation. In short, *Monopoly Capital* unconvincingly uses isolated periods of economic history to "demonstrate" the existence of "inexorable" forces.

LIMITATIONS of space prohibit a detailed discussion of the general social effects Baran and Sweezy attribute to monopoly capitalism: the racial crisis, decaying cities, traffic congestion, backward educational systems, etc. Clearly a great deal is wrong in the United States and irrationality and injustice abound. But it will not do simply to attribute it all to the power of "monopoly capitalism." If the corporate elite really had a substantial economic interest in the maintenance and worsening of urban and racial problems, it would be the most suicidal elite that ever lived.

I am not saying that our social problems are transitory or that the American system of mixed capitalism is basically "good," requiring only some minor changes. Rather, Baran and Sweezy, with their total emphasis on the corporate oligarchies, have failed to provide an adequate economic analysis of American society. If there is one possible from a Marxist point of view, it remains to be done.

INDEX

Index